IT'S
A
MIRACLE!?

IT'S A MIRACLE!?

What Modern Science Tells Us about Popular Bible Stories

CLAY FARRIS NAFF

PITCHSTONE PUBLISHING
DURHAM, NORTH CAROLINA

Pitchstone Publishing
Durham, North Carolina
www.pitchstonepublishing.com
Copyright © 2018 by Clay Farris Naff

10 9 8 7 6 5 4 3 2 1

Library of Congress Cataloging-in-Publication Data

Names: Naff, Clay Farris, author.
Title: It's a miracle!? : what modern science tells us about popular Bible
   stories / Clay Farris Naff.
Description: Durham, North Carolina : Pitchstone Publishing, 2018.
Identifiers: LCCN 2017058149 | ISBN 9781634311540 (pbk. : alk. paper) | ISBN
   9781634311564 (epdf) | ISBN 9781634311571 (mobi)
Subjects: LCSH: Religion—Controversial literature. | Religion and science. |
   Miracles—Biblical teaching. | Bible stories.
Classification: LCC BL2776 .N34 2018 | DDC 220.6—dc23
LC record available at https://lccn.loc.gov/2017058149

CONTENTS

INTRODUCTION

 What's Science Got to Do with It?

Why would anyone presume to apply science to Bible stories? Isn't it enough to say that by faith alone we know they are true? Or to say that we're sure they are nothing more than folklore? In this book, we're going to do something amazing: we'll use the tools of science to both debunk and, in at least some sense, validate miracles of the Bible—occasionally doing both for the same story. Will that mean we know the Bible's stories are either true or false? To address that question, we have to think about what it means to know something. But fair warning: if egghead arguments burn your bacon, this here will be the boring part of the book for you. Feel free to skip to chapter 1.

Philosophers in a line stretching back to the ancient Greeks tell us that knowledge is justified true belief. Trust me, we need both modifiers. A person who holds a true belief without justification is lucky rather than knowledgeable. I might truly believe that the horse whose name is on the betting stub I hold will win the Preakness. My pick—let's call him Plumber's Helper—may be a mighty fine colt, but until the race is run I cannot claim to *know* who the winner will be.

"Justified true belief" remains a pretty useful definition of knowledge, but—full disclosure!—for the past century some philosophers have disputed that claim. It's easy enough to prove that a specific belief is false—say, the claim that Abraham Lincoln served three terms as president and built the Panama Canal—but is it ever possible to prove that a general claim is true or false? Even the most plodding, sure-footed facts can slip. Will the sun rise tomorrow? Quite likely it will, but astronomers tell us, on good evidence, that a day will come when it won't. Instead, a billowing, reddening sun will swallow up the Earth. But who knows? Maybe we'll have moved on by then.

Anyway, we don't have to wait a billion years to witness knowledge crumbling. What was widely thought to be justified and true at one time often becomes widely discredited at another. For example, in the 1930s a majority of doctors smoked, and buffered by reassuring advertisements featuring doctors, the public believed that cigarettes were healthful. By the 1960s that apparently justified belief began to swing around. A few decades later it was entirely snuffed out.

Today, there is no universally accepted philosophical definition of knowledge. Postmodernists argue that truth is merely an expression of power. Something is true because I say so. Or an authorized text says so. Or maybe because the "dominant discourse" says so. Or maybe nothing is true.

Richard Rorty (1931–2007), an influential American philosopher, championed a "conversationalist view of truth and knowledge." Fond of putting things provocatively, he once said, "Truth is what your contemporaries let you get away with."

This, of course, is nonsense. Rulers from Caligula to North Korea's Kim Il-sung have claimed to be gods, and moreover have had the power to force that view on the populace, yet not one of them has cheated death.

Here, then, is our dilemma: we often find it convenient, attractive even, to dispute the claims of experts by claiming that knowledge is a social or political construct. British politician Michael Gove exploited that craving as a leader of the successful "Brexit" campaign. Asked to name an economist who favored his view that Britain should vote to withdraw from the European Union, he refused, saying, "People in this country have had enough of experts."

Of course, in an era of "post-truth politics," to borrow blogger Dave Roberts' phrase, some politicians will say anything. But even scientists, who have the keenest respect for data, are leery of facts and mostly reject the idea of a discoverable, capital "T" Truth out there. Sean Carroll, a thoughtful but far-from-squishy theoretical physicist, observes, "We use words like 'data' or 'evidence,' but the concept of a 'fact' simply isn't that useful in scientific practice." And to, ahem, *prove* the point he gathers some data. His search through titles in a high-energy physics database of papers yields 9,909 instances of "data," 4,396 hits of "evidence," but a paltry 120 uses of "fact."

In short, scientists generally admit that all they have in hand are probabilities and approximate models of reality. Law of gravity? Sure, but as Einstein showed, things could change tomorrow. His general law of relativity (GR) showed that Newton's law was an incomplete description of gravity (not to mention time and space). No one thinks that GR is the last word on reality. More important, the fundamental regularities of the Universe, as observed in exquisite detail by 500 years of modern science, are in principle subject to change. The laws of nature have proven so reliable that NASA can send a probe to Pluto and your GPS will guide you to the nearest gas station. Yet, gravity might evolve. The sun might not come up tomorrow.

On the other hand, we long for certainty. We feel sure that we know some things. But if asked, "How do you know?" we resort to radically different grounds to justify our knowledge. Nowhere do the grounds differ more than in science and religion.

To take the last first, religious knowledge is often said to be revealed. That is to say, it passes from the mind of God to a messenger, prophet, or scribe and into your mind. Occasionally, God skips the middleman and speaks directly to a believer. According to T. M. Luhrmann, a distinguished professor of psychological anthropology at Stanford University, nearly a quarter of Americans claim to hear the voice of God—far more than could be explained away by mental illness, she says.

As knowledge, however, the voice of God is problematic. For one thing, the messages people claim to receive are generally brief, vague, and unverifiable. Luhrmann reports that Martin Luther King Jr. heard Jesus whisper, "I will be with you." True? False? True one day but not the next? Who can say?

When God does get specific, it's often in such horrifying ways as to invite disbelief. On Mother's Day in 2003, churchgoer Deanna Laney of New Chapel Hill, Texas, bashed in her sons' heads with rocks. Two of the three died. Laney said God had ordered her to kill them as a test of her faith. A jury found her not guilty by reason of insanity.

That seems a reasonable verdict, but it raises some troubling questions about religious knowledge. How can a person who hears God's voice know if it's authentic or not? If Laney was a Bible reader, she would have found plenty of confirmatory evidence: God orders Abraham to kill his son as a test of faith, then halts the test at the last possible moment. God kills all the firstborn sons in Egypt. God slaughters some Israelite children who poke fun at one of his prophets. So, why shouldn't a housewife in Texas believe that God wants her to kill her own sons? For that matter, why shouldn't a jury believe it?

One reason: everyone knows that some people "hear" voices in their heads that are not God. In the Bible, they are described as demons; in modern psychiatric parlance they are auditory hallucinations. If God does occasionally speak to some people, it's odd that he doesn't claim exclusive use of the channel. He surely has the power to prevent any stray messages from clogging up the brainwaves.

It's not just a matter of screening out psychosis. People all over the world claim to experience mystical revelations, but the content is nearly always in keeping with their cultures. In India, Hindus actively pursue direct revelation as part of religious practice. When it happens what they see is not the face of God but *Brahman*, a metaphysical reality described as pervasive, genderless, infinite, eternal, true bliss. Not exactly the old man with the flowing white beard found on the ceiling of the Sistine Chapel, eh?

More troubling, it appears that mystical experience can be induced in even the most recalcitrant subject. Philosopher of religion Tim Holt observes, "Using the transcranial magnetic stimulator to apply a magnetic field to the temporal lobes can cause people to experience God. This phenomenon is not limited only to believers; even atheists can be caused to have religious experiences using the transcranial magnetic stimulator."

Given the variety of direct religious experience, is it possible to single out an authentic revelation, if any, from delusion, fakery, mass hysteria, culturally prompted experience, or medically induced hallucination? God knows.

In practice, religious people typically acquire a specific set of beliefs through indoctrination, usually beginning in childhood. An authority figure such as a priest or a "Sunday School" teacher reads sacred texts to them and interprets their meaning.

Can transmitted belief be considered knowledge? Seen from within any particular religion, it surely can. In Protestant Christianity, for example, we find the concept of *sola fide*—by faith alone. It's the idea that simply believing the Bible's claims justifies a person's salvation—assuming God goes along. But is this true? How can we know? There are a great many religious narratives in the world: Can they all be true? Can a religion with internal contradictions be true?

These are perplexing questions, but at this point we must face a mundane fact: most religious people, when asked in surveys why they believe in God, don't cite faith or Scripture; they cite evidence. Despite overwhelming scientific evidence to the contrary, many religious people perceive the world as exhibiting design and purpose. Perhaps this is natural. Science is now gathering a growing body of evidence that such perceptions, and the beliefs they underpin, are hard-wired into each of us—to a greater or lesser degree.

Some claims may be overstated—it's not clear that there's a "God" gene, and the "God part of the brain" may only be metaphor, but for sure we are born with a tendency to see purpose and meaning in events. Psychologist Deborah Keleman of Boston University's Child Cognition Lab has done research showing that when children give explanations for natural phenomena, they go for purposeful explanations. This, she says, is much like what we find in religions around the world, "this idea that things exist for some purpose, because of some agent's purpose."

And no wonder! For nearly all of history it seemed entirely reasonable to carry on thinking that way for a lifetime. Yet, science has changed that. For a growing number of people, the balance of evidence tips against design and purpose. It's this clash between a widespread belief in the Bible as a valid description of the world and a growing challenge from those who read the evidence of the world around us in a profoundly diferent way that leads us to abandon the pursuit of religious knowledge, and turn instead to the one kind of knowledge that transcends all cultures: science.

So, we are going to apply scientific methods of inquiry, explanation, and, yes, speculation to the sprawling narrative known as the Bible—

more particularly, to many of the miracles appearing in the tale. That means we need to know what counts as scientific knowledge, and what counts as a miracle. To take the last first, the *Stanford Dictionary of Philosophy* tells us that a miracle is "an event that is not explicable by natural causes alone." Taken at face value, that would seem to deflate the tires on this project before it moves out of the driveway. However, we don't have to give up yet.

There are several interesting possibilities. One is that there is an overlooked natural explanation for the event, and that people have been fooled or have deceived themselves into accepting a supernatural explanation. For the most part, we will focus on such circumstances. Of course, they will be inadequate to explain every aspect of the biblical miracles we examine. But, for sake of better understanding, it is worth teasing out the core of the story—the strange event—and assuming that the rest is embroidery or apologetics (rationales intended to bolster a religious claim).

These are not wild assumptions. The tendency to embroider is as natural as storytelling itself. Every culture has its tall tales and legends, and ours is no exception. From Paul Bunyan to Bigfoot and from the Bermuda Triangle to Area 51, the brief span of American history is replete with mysterious and wildly exaggerated tales. Other cultures offer myths as comprehensive as those from the Bible. They include creation stories utterly different from the Bible's twin accounts, such as the magic reed story of the Navajo, the Shinto island-creation story, and the Hindu story of cyclical creation. So, peering inside biblical miracles for a natural event that may have prompted the story is an entirely reasonable undertaking.

A second interesting possibility is that some miracles were simply frauds—products of the illusionist's art, presented as magic. Again, we know that illusionists have been a part of human culture for millennia, and we know that even today they sometimes commit frauds on the gullible. The illusionist known as The Amazing Randi has dedicated a large part of his career to unmasking such frauds. But there was no Amazing Randi in the ancient Near East, and no one had a grasp of inviolable scientific principles that give rise to skepticism about extraordinary claims. Indeed, even today if you ask anyone who believes in ghosts about the conservation laws of physics, you invariably draw blank stares. No one who understands those laws, their basis, and the overwhelming evidence that supports them spends any time worrying about ghosts.

A third possibility is that the Bible's miracle stories are literal accounts of actual happenings. Apologists exert a great deal of effort to promote this possibility, but all too often they get the science wrong or invoke magic. That's too bad. Magic is actually the most boring of all explanations. "God willed it" tells us nothing of interest. Miracles aren't even instructive. No one seeking to find moral guidance from God's miracles can avoid picking and choosing among their implications. (See bears-children below.) Ambiguity in Scripture may be why you find Christians passionately for and against the death penalty, for example. (Incidentally, Muslims face the same dilemma. Some read the Quran as an injunction for mercy and compassion; others see it as a license for murder.) But what if God chose to perform miracles via natural means? Or plausible but as yet untested natural means, such as teleporting technology back through time? Now that could make for some interesting scenarios!

Showing *how* leaves a deeper point unaddressed: if a miracle is an event that defies normal explanation, then *why* attempt to use science to explain it? Fair question. Fortunately, for the sake of reader and writer alike, there are at least three good answers.

First, if miracles are, as many believers claim, signs from God, then it is critically important to distinguish the authentic from the phony. If science can provide a plausible explanation for an event, then it's not truly a miracle. For example, a sighting of the Madonna is a commonly claimed miracle. Since we now know things the ancients did not—that hallucinations can be prompted by many things, from sleep or oxygen deprivation to peyote to a powerful magnetic field, and that moreover group delusion is easy to induce—should we include such claims in the roster of miracles? If God wanted to send a sign, mightn't we expect one that is unmistakable? If dispatched by the deity, shouldn't the Madonna appear, say, as a guest on the *Late Show* on CBS rather than as a ghostly apparition perched in a tree in Cameroon?

Second, it is epistemically important to maintain the coherence of science by explaining away mysteries in natural terms—where we can. Science is only as reliable as nature is consistent. If spectacular miracles occur, why not trivial ones? Indeed, sporting events often give rise to claims of a minor miracle, with winning players giving credit for their victories to God. But if the deity chooses winners of football games, soccer matches, and horse races, how can we assume that he does not alter the results of pharmaceutical trials to favor one drug over another? Indeed, what scientific result could be considered reliable in a world where any sort of miracle takes place?

Third, if miracles are authentic, it is theologically important to learn via analysis what kind of deity stands behind them. If God created both the Universe we inhabit and the laws that govern them, then how and why would he violate those laws? Indeed, can a supremely perfect God be a law-breaker? If not supremely perfect, what's he like?

For example, in 2 Kings 2:23–24, the prophet Elisha, who has been wandering about the Holy Land invoking miracles for pay, approaches the gates of Bethel. Before he can enter, a large group of children comes pouring out and mocks him for his bald head. Elisha curses them and, hey, presto, a pair of bears comes loping along and mauls 42 children to death. It's hard to see why anyone would want to claim this as a miracle on behalf of a just, loving, and perfect God. Not even in Texas does corporal punishment go that far.

In any event, a ready explanation springs to mind: coincidence. There are bears in the Middle East. They are Syrian brown bears, which happen to be among the smaller and shyer of the species. They live in the mountains, not in the plains, and they rarely attack humans. But even shy bears will wander if food runs short, and hungry or nervous bears may attack—especially if their young are nearby.

It's notable, then, that the Bible describes the attackers as "she-bears." A group of youngsters calling out, "Ha, ha, check out old baldy!" from the wooded side of the road might provoke bears to attack. What's utterly implausible, however, is that two bears could maul 42 children. Even if the kids had been paralyzed with fear—not a likely conjecture—the bears surely would have lost interest less than halfway through the job. On the whole, Syrian brown bears would rather snack on pine nuts than brats. So, along with coincidence, to explain this we have to include a large dollop of exaggeration.

In suchlike fashion, we will explore the miracles recorded in the Bible. Not every miracle, of course. Some are trivial, some are repetitive, and some are incoherent. But most, like the unfortunate boys of Bethel, are fair game.

Speaking of fair game, though, before we go further we must specify what counts as science. For starters, must it be natural? The consensus view is that it must, although a lively argument exists as to whether this expresses a worldview—naturalism—or is merely a method of science.

This problem existed long before the emergence of modern science. People everywhere have a naive and largely instinctive sense of what is natural (as well as an agency instinct that promotes belief in the supernatural). As we will see, the ancients who wrote the Bible sometimes described as miracles things that we take to be quite natural. Earthquakes, for instance, can be much more easily and consistently explained by plate tectonics than God's wrath. But there are plenty of other stories in the Bible that defy easy scientific explanation. That's where the fun comes in.

So what is a scientific explanation? Dictionary definitions of science are generally dry, abstract, and deeply inadequate. Here, for example, is Merriam-Webster's primary definition: "Science 1: knowledge about the natural world that is based on facts learned through experiments and observation." What most of us learned in the public school classroom is hardly any better. This is changing for the better, but in too many instances American pedagogy ranges from the good old science-fair model (observation-hypothesis-experiment-conclusion) to "shut up and learn these facts for the quiz on Friday."

As a science writer I've had many opportunities to watch scientists at work and to listen to them think out loud. I know that Nobel Prize–winning physicist Max Born was right when he said: "There is no philosophical high-road in science. . . . No, we are in a jungle and find our way by trial and error, building our roads behind us as we proceed." Yet, as it hacks through the jungle of the unknown, science leaves in its trail a lawful, coherent, and reliable description of the world.

By lawful I mean that certain simple statements about the world prove reliable again and again, whether put to the test in one culture or another, or at one time or another. An example you can rely on in all but the most extreme conditions is this: with any two magnets, like poles repel and unlike poles attract. It may yet be that under some never-yet-seen conditions, there is a bizarre phenomenon called a magnetic monopole, but even if this proves to be the case, in ordinary conditions the usual rules will still apply. You can rely on that.

By coherent, I mean that one area of science meshes with all the other established areas of science. Roughly speaking, there is a nested hierarchy of sciences. Physics is the most fundamental, dealing with the elementary particles (or fields) and forces of nature. Every other science must, therefore, cohere with physics. If a conflict arises, it may mean that an unknown mechanism of reconciliation has yet to be discovered (as in the quarrelsome cousins, general relativity and quantum physics), or it may mean that we are faced with a pseudoscience.

Astrology is a stellar example of the latter. We know this not only because reading the stars fails to consistently predict human affairs, but also because the stars are simply too distant to coordinate their influence. The stars of the Big Dipper, that most familiar of constellations to those of us in the West, are not only far too distant from us to have real-time effects. They are also far too distant from each other. They may appear to lie together on the inner surface of a dome, but in reality the closest is 63 light-years away, while the furthest is 210 light-years distant. Just to coordinate a single event would take the stars of the Big Dipper a minimum of 147 years, plus another 63 to deliver the message to Earth. Anything less would violate a fundamental law of physics: information cannot travel faster than the speed of light. Unlike the fussy rule about ending a sentence with a preposition, the speed-of-light limit is a law ~~you can rely on~~ upon which you can rely.

By reliable I do not mean that science is guaranteed to be The Truth, but that it provides the best available explanation of a given observation that is consistent with the evidence. Tomorrow, better data or better interpretation may give us an even better explanation. In most cases that won't render the old explanation useless; it will simply give us a more refined picture of nature. A nice example is Newton's law of gravity. In 1687, he described it as an attractive force between any two massive objects that is directly proportional to the product of their masses and inversely proportional to the square of the distance between them. Put as an equation, it works exquisitely well to predict the behavior of objects. Just ask NASA, which uses Newtonian methods to plot the paths of its planetary probes.

But Newton didn't have the last word. In 1915, Einstein published his general theory of relativity, which describes gravity as a warping of space-time that bends the path of objects, and in 1919 observations of starlight bent by the sun validated the theory. So, Newton is wrong. But so is Einstein. Neither one offers a perfect description of gravity. Instead, Newton's laws work just fine for nearly every human endeavor. When extremely high speeds and/or strong gravitational forces are involved, we have to make use of Einstein's more accurate equations. When gravity becomes so extreme that a black hole forms, Einstein's equations stumble over quantum mechanics.

One more characteristic needs to be taken into account. Science is more than laws or equations. It is also a story that explains the natural world. The explanatory side of science can be found in theory. It is an irritating quirk of the English language that in popular usage the word "theory" means a hunch. A hunch is like Sherlock Holmes's best guess about who killed the lord of the manor. In science "theory" means something quite different. It is the end of the road, after the jungle has been cleared. Theory gathers all the available data about a particular phenomenon and embeds them in a consistent explanatory tale. Germ theory, for example, is not a wild guess about what makes people sick (look up the history of chiropractic for that), but rather a comprehensively tested explanation for how microbes cause infectious disease.

With all this in mind, what counts as a scientific explanation in this book is a tale that fits the facts—or a reasonable restatement of the facts—and is consistent with the relevant branches of science. These will range from physics, astronomy, and geology to biology, psychology, and neuroscience. As for kinds of explanation, you can look forward to weather, volcanism, cometary impact, chance, and a host of other natural phenomena, including human frailties such as delusion, misinterpretation, beguilement, and confabulation. Just to stretch the boundaries of the possible, we'll consider what would happen if God or his surrogate were all knowing but constrained to perform miracles by natural means. Borrowing from fantasy novelist Terry Pratchett, we'll call this kind of explanation "technomancy." We'll indulge in some science-based speculation as well.

Will all those explanations be true? Will any? Not necessarily. Just because something can be described in natural terms consistent with science doesn't mean it's true. Claiming that our universe is a simulation running on a computer is (so far) consistent with science. That doesn't make it necessarily true.

What's more, while the findings of science make an airtight case for naturalism, that in itself does not rule out the involvement of an external agency. Say, you buy a lottery ticket and pray that your number wins. The next day you check and find that you have indeed won millions. How can we explain this? It might be mere chance. Nothing unnatural about that. Even though the odds against scooping the jackpot are typically close to 300 million to 1, with enough tickets in circulation a winner will come up, just by chance. But we cannot prove thereby that God never puts a finger on the spinning wheel. The deeper question is this: why would God listen to your prayer and brush aside those of all the other poor believers who bought a ticket?

If you allow God the power to be arbitrary, unjust, or unintelligible, then anything logically possible may indeed be. Curiously enough, science has its counterpart: in quantum physics there arises what is known as the totalitarian principle—whatever is not forbidden is mandatory, or to put it a bit more elegantly, anything that can be must be. With all the foregoing in mind, read on at ease, knowing that nothing in the pages that follow can either destroy a well-founded faith or turn skepticism into credulity. But I hope you will learn a thing or two, have a little fun, and stretch your imagination.

* * *

A word on biblical text: There are many versions of the Bible. In this book, we draw on just two: the New Revised Standard Version (NRSV), released in 1989, for accuracy, and for poetic beauty the King James Version (KJV), compiled in the early 1600s, where the text is substantially the same as that which modern scholarship has placed in the NRSV.

1

Creation Stories

"When we read about Creation in Genesis, we run the risk of imagining God was a magician, with a magic wand able to do everything. But that is not so."

—Pope Francis, October 2014

Perhaps the most surprising fact about the origin stories of the Bible and science is what they have in common. No, I'm not referring to the Big Bang. It's spiffy that Pope Francis, along with many fellow Catholics and most mainstream Protestants, regards the Big Bang theory as authentic, but a literal mind will find it hard to reconcile the science of cosmogony (cosmic origins) with this:

> In the beginning when God created the heavens and the earth, the earth was a formless void and darkness covered the face of the deep, while a wind from God swept over the face of the waters. Then God said, "Let there be light"; and there was light. (Genesis 1:1–3, NRSV)

Wind? Waters? Both of those require oxygen, but it would be billions of years, science tells us, before any oxygen atoms were forged in the first generation of stars and then liberated in supernovae to roam the heavens and pair up with hydrogen.

Not even light could travel freely until long after the Big Bang. What we detect now as the cosmic microwave background radiation was liberated only when the Universe grew large enough to give photons some elbow room in the boiling particle soup. Cosmologists tell us that the Dark Ages that followed the Big Bang lasted some 400,000 years. Yet, Genesis 1 goes on to tell us:

> And God saw that the light was good; and God separated the light from the darkness. God called the light Day, and the darkness he called Night. And there was evening and there was morning, the first day.

Day? What's a day before there's a Sun or Earth? Does God have a celestial grandfather clock? If so, it must ticktock mighty slow. For, according to the evidence-based scientific narrative, two generations of stars would rise and fall and nine billion years would drift by before the Earth was born.

The discrepancies only widen as we go on. Genesis 1 continues:

> And God said, "Let the waters under the sky be gathered together into one place, and let the dry land appear." And it was so. God called the dry land Earth, and the waters that were gathered together he called Seas. And God saw that it was good. Then God said, "Let the earth put forth vegetation: plants yielding seed, and fruit trees of every kind on earth that bear fruit with the seed in it." And it was so. The earth brought forth vegetation: plants yielding seed of every kind, and trees of every kind bearing fruit with the seed in it. And God saw that it was good. And there was evening and there was morning, the third day.
>
> And God said, "Let there be lights in the dome of the sky to separate the day from the night; and let them be for signs and for seasons and for days and years, and let them be lights in the dome of the sky to give light upon the earth." And it was so. God made the two great lights—the greater light to rule the day and the lesser light to rule the night—and the stars. God set them in the dome of the sky to give light upon the earth, to rule over the day and over the night, and to separate the light from the darkness. And God saw that it was good. And there was evening and there was morning, the fourth day. (Genesis 1: 9–19, NRSV)

Wait. A dome? You mean, like in a planetarium? That doesn't match up too well with our picture of the universe today. As you can readily imagine in the mind's eye, stars in a dome would all be about the same distance from an observer on Earth. What's more, set in a dome, no star could lie before or behind another. But starting in 1838, astronomers observed stellar parallax—the apparent shift of nearby stars in relation to much more distant stars as the Earth trundled around the Sun. As even nearby stars are immensely distant, parallax is hard to measure, but with better telescopes and recording methods, any clinging beliefs that stars lie in a dome vanished. A vast, dark ocean of stars gradually came to represent our sketch of reality.

By the early twentieth century, it became evident that stars gather in gravitationally bound blobs and swirls called galaxies. In 1929, astronomer Edwin Hubble showed that, like raisins in a rising loaf of bread, the galaxies are moving away from one another. The Milky Way, our home state, so to speak, contains about 100 billion stars—more than 10 times the number of people on Earth.

The Hubble Telescope and its space-based successors have shown that the number of galaxies is immense. By the most recent reckoning, there are at least 2 trillion galaxies in the visible universe. If ours is typical, that presses the number of stars beyond counting, let alone naming.

The contemporary scientific account dates the "creation" of the Universe to about 13.8 billion years ago. According to the most widely accepted interpretation of the evidence, an event called the "Big Bang" allowed an incomprehensibly tiny and dense dot of positive and negative energy to expand at breakneck speed. As it did, time and space took shape, and the fundamental forces of nature—gravity, electromagnetism, and the atomic strong and weak forces—shook out of the protoforce and largely dictated the course of events thereafter.

None of this matches up well with the biblical account. The idea that a great light (the Sun) and a lesser light (the Moon) separate the light from the dark on a universal scale now looks laughable. The Sun isn't even the most impressive bulb in the stellar string decorating our spiral arm of the Milky Way.

What on Earth (so to speak) do the biblical and scientific accounts have in common? They each come in multiple, contradictory versions.

So far, we've just dipped a toe into the waters of Genesis 1. But there's another account of creation in the Bible. Genesis 2 offers a version that differs in some key details from Genesis 1.

Scholars have determined that the two versions had differing sources: Genesis 1 has its roots in a Babylonian origin tale of the god Marduk (who gets several mentions in the Bible). It appears that when the Jews were forcibly moved to Babylon, after the conquest and sacking of their temple around 586 BCE, they took their long-held creation story but were confronted with another. Eventually the two were uneasily stitched together. Here's a snippet of an excellent account by law professor Doug Linder of the University of Missouri–Kansas City:

> In the beginning, about 3,000 years ago, Jewish desert dwellers in what is present-day southern Israel told a story around campfires about the creation of the first man and first woman. . . . Everything goes well for a spell, in the story told in the desert, but then the Creator is disobeyed and bad things start to happen.
>
> Four or five centuries later, five-hundred-plus miles to the east in what is most likely present-day Iraq, a remarkable Jewish writer—whose name we do not know—set about the ambitious task of constructing a primary history of his people. Evil Merodach [Marduk] reigned in this dark time of Jewish exile, around 560 B.C., and the writer hoped that his history would help his people endure their many trials. The writer was most likely a priest, and might have been assisted in his work by other priests and scribes. . . .
>
> The priest wove the two texts together, trying to avoid repetition and altering them where necessary to avoid blatant inconsistencies. The priest confronted an additional problem: the two texts originally reflected views about two different gods in a time of polytheism, but by the time he compiled his history, belief in a single god had become prevalent among Jews. The priest, therefore, sought to remove passages supporting the polytheism of an earlier age—and, except for a few hints here and there, he succeeded. Finally, he added some writing of his own, or of his priestly contemporaries, that reflected the ideas of his own, more mature, period of Judaism.

What other differences remain? In Genesis 1, animals come first, then humans:

And God said, "Let the earth bring forth living creatures of every kind: cattle and creeping things and wild animals of the earth of every kind." And it was so. . . . Then God said, "Let us make humankind in our image, according to our likeness . . ." (Genesis 1: 24–26, NRSV)

In Genesis 2, it's the other way around. Adam, at least, comes first:

In the day that the Lord God made the earth and the heavens, when no plant of the field was yet in the earth and no herb of the field had yet sprung up . . . the Lord God formed man from the dust of the ground, and breathed into his nostrils the breath of life; and the man became a living being. . . . Then the Lord God said, "It is not good that the man should be alone; I will make him a helper as his partner." So out of the ground the Lord God formed every animal of the field and every bird of the air, and brought them to the man to see what he would call them. (Genesis 2:4–19, NRSV)

Oh, and let's note that in Genesis 1, Adam and Eve are created together, but in Genesis 2, Eve is an afterthought when the animals fail to meet the, uh, social needs of Adam:

The man gave names to all cattle, and to the birds of the air, and to every animal of the field; but for the man there was not found a helper as his partner. So the Lord God caused a deep sleep to fall upon the man, and he slept; then he took one of his ribs and closed up its place with flesh. And the rib that the Lord God had taken from the man he made into a woman and brought her to the man. (Genesis 2:20–22, NRSV)

Apologists may bend themselves into pretzels trying to explain away the inconsistencies between Genesis 1 and 2, but at least science has its story straight . . . right?

Well not exactly. Although the Big Bang Theory is so well accepted that it can serve as the title of a popular sitcom, there are some deep problems in the narrative. Actually, somewhat like Genesis, there are two versions of the contemporary narrative. The smug version claims that all the important problems have been cleared up, and that we have a (virtually) complete and reliable explanation. This version is typified by physicist Lawrence Krauss's 2012 trade book, *A Universe from Nothing: Why There Is Something Rather than Nothing.*

The book attempts to refute a claim made by Christian philosophers that the existence of the universe necessitates the existence of God. It's a claim that draws on the medieval Christian philosopher Thomas Aquinas, who in turn drew from Aristotle. The entire argument, Krauss points out, rests on the false premise that everything has a cause, except the First Cause, which is God. (Uncaused things, such as the random decay of radioactive atoms, happen all the time.) Unfortunately, Krauss's book rests on another false premise: that the laws of quantum physics governing our Universe count as nothing. In a blistering review, philosopher David Albert notes that "vacuum states—no less than giraffes or refrigerators or solar systems—are particular arrangements of elementary physical stuff."

> And if what we formerly took for nothing turns out, on closer examination, to have the makings of protons and neutrons and tables and chairs and planets and solar systems and galaxies and universes in it, then it wasn't nothing, and it couldn't have been nothing, in the first place. . . . The whole business of approaching the struggle with religion as if it were a card game, or a horse race, or some kind of battle of wits, just feels all wrong.

Ouch.

The strongest confirmation of the Big Bang is the cosmic microwave background radiation (CMB). It is also the source of a confounding mystery. Two mysteries, actually. The first concerns why it's so bland, and the second why it's so bearable. The answer to each is a scientific bugbear.

We begin with the bland: In 1964, a pair of young American radio astronomers named Arno Penzias and Robert Woodrow Wilson were setting up a large receiver at Bell Labs in New Jersey. As they tried to tune into the stars, they kept getting an annoying hiss. They tried everything to eliminate it, including scraping away pigeon droppings by hand. At last they realized that they had stumbled onto something magnificent:

the first light of the universe. Known now as the cosmic background radiation, these are photons that broke free from the swelling fireball that was our early universe the moment that empty space first appeared between the jam-packed elementary particles, some 400,000 years after the Big Bang. Russian-born theoretical physicist George Gamow had predicted the CMB nearly two decades earlier, but the poor guy got left in the cold when the Nobel prizes were handed out.

In 1995, the first in a series of satellites took images of the CMB. One image, taken by a probe called W-MAP, found variations in the CMB that allowed scientists to calculate the age of the universe with a high degree of confidence. So far, so good. But the variations they found were only visible because they had been tremendously enhanced. Actually, the CMB from any direction varies by no more than 1 part in 10,000—maybe in 100,000—and then not by much. Think of pouring a ten-pound sack of rice onto a table, and trying to find the one slightly gray grain among all the white ones. It's an almost superhuman task; you'd surely need a computer or a graduate student to get it done.

Given the age and size of the universe, that's a mystery, because like wind blowing off the Arctic and wind blowing in from the Sahara, you'd expect that one would be cold and the other hot—at least until they'd had a chance to mix. Instead, our great big universe seems to be incredibly bland.

There's an answer to that: cosmic inflation. In the early 1980s, Alan Guth had the brilliant idea that maybe the early universe puffed up suddenly, like a kernel in hot oil blowing up into popcorn—only much, much faster. His idea, grounded in a complex tangle of technical detail, answered many questions about how the universe looks today, including its blandness. Better yet, it has a plausible mechanism: a tiny, random quantum fluctuation way back in the beginning. And, holy mackerel, it even fits together with the solution to another big problem: why is the Universe livable?

Now don't get me wrong: most of the Universe is overwhelmingly hostile to life. Unprotected, you wouldn't last a minute in space, and space is pretty much all there is. On at least one of the tiny dots that constitute habitable zones, however, we know that intelligent life has sprung up—well, occasionally intelligent, anyhow.

But theoretical calculations say that should not be. Dark energy is the most powerful force in the Universe. Discovered in the 1990s, it is estimated to constitute about two-thirds of all the energy there is, and it has but one mission: to drive everything apart from everything else. Dark energy is causing galaxies to race away from one another at an ever-accelerating rate. But here's what makes it a dark mystery: every attempt to calculate what dark energy should be, given what we know about the laws of the Universe, leads to the conclusion that it should be vastly more powerful that it is. About 120 orders of magnitude more powerful. If it were anywhere near the theoretical value, no stars would ever have formed. The Universe would have just blown up like one of George Lucas's Imperial Death Stars, and that would be that.

So, how to explain it? The only way, concluded Nobel laureate Steven Weinberg and many other physicists, is to assume that the quantum jiggle that triggered cosmic inflation for our Universe also happened in lots of universes, each with a random set of parameters. This is known as the multiverse conjecture. Think of bubbles in club soda. The bubble that we're in just happens to have parameters that allow for humans, and more specifically, for cosmologists to grow up in so that they can imagine the answer to the question of how we got here in a peer-reviewed scientific paper. It's called the anthropic principle.

This explanation dovetails nicely with cosmic inflation, which in principle can make lots of bubbles of various configurations. So, for a couple of decades, cosmic inflation became the presumed heir to the explanatory throne.

But then, just as the Bible's creation story becomes self-contradictory, this tale turns on itself. Scientists gradually realized that if the anthropic principle explains the value of dark energy, it can explain the value of any other force in nature. But if all those values are up for grabs, then in principle dark energy could have been stronger than it is, if only gravity were also stronger. And so on. So much for explaining, and more important, predicting. If an idea can't furnish a testable prediction, it's generally not considered scientific.

Worse yet, as one of the developers of cosmic inflation theory realized, in opening the quantum cage to allow inflation to take wing, theoreticians had unleashed a monster: eternal inflation. The clear implication of their calculations is that if our universe started that way, then all around us eternal inflation is raging, with a few islands of calm, and even fewer with complex order. The whole shebang is ruled by the totalitarian principle: anything that can happen will happen—over and over again. Infinity's a bitch.

Having helped to set all this in motion, Paul Steinhardt now says: "To me, the accidental universe idea is scientifically meaningless because it explains nothing and predicts nothing." Beside, it may not be true. The objections to inflation theory and its implications are mounting as new data come in from space-based observatories. Steinhardt and colleagues are working on an alternative scenario—a kind of Genesis 2.

Any way you look at it, the biblical creation story and science remain parsecs apart. Can they possibly be reconciled? To put it another way, could any creation story fit with science? That's a question we will return to at the end of this book.

2

 ## *Water, Water, Everywhere:*
The Great Flood

Ancient tales of a great deluge abound, but there's none quite so influential as the Noachian Flood. People still produce photos of rills and humps on Mount Ararat that they breathlessly proclaim to be the remains of Noah's Ark. A creationist museum in Kentucky has built what it claims to be a reproduction of the original seagoing zoo. What was the Ark *really* like? Here's how the Bible describes it:

> Noah was six hundred years old when the flood of waters came on the earth. And Noah with his sons and his wife and his sons' wives went into the ark to escape the waters of the flood. Of clean animals, and of animals that are not clean, and of birds, and of everything that creeps on the ground, two and two, male and female, went into the ark with Noah, as God had commanded Noah. And after seven days the waters of the flood came on the earth.
>
> In the six hundredth year of Noah's life, in the second month, on the seventeenth day of the month, on that day all the fountains of the great deep burst forth, and the windows of the heavens were opened. The rain fell on the earth forty days and forty nights. On the very same day Noah with his sons, Shem and Ham and Japheth, and Noah's wife and the three wives of his sons entered the ark, they and every wild animal of every kind, and all domestic animals of every kind, and every creeping thing that creeps on the earth, and every bird of every kind—every bird, every winged creature. They went into the

ark with Noah, two and two of all flesh in which there was the breath of life. And those that entered, male and female of all flesh, went in as God had commanded him; and the Lord shut him in. (Genesis 7:6–16, NRSV)

No wonder the Lord slammed the cabin door, for by now the ark must stink to high heaven. Things can only get worse as millions of corpses bob in the waters. Perhaps to tamp down the stench, God keeps the spigots open for months.

The waters swelled so mightily on the earth that all the high mountains under the whole heaven were covered; the waters swelled above the mountains, covering them fifteen cubits deep. And all flesh died that moved on the earth, birds, domestic animals, wild animals, all swarming creatures that swarm on the earth, and all human beings; everything on dry land in whose nostrils was the breath of life died. He blotted out every living thing that was on the face of the ground, human beings and animals and creeping things and birds of the air; they were blotted out from the earth. Only Noah was left, and those that were with him in the ark. And the waters swelled on the earth for one hundred fifty days. (Genesis 7: 19–24, NRSV)

Okay, so now the entire planet is covered in water to a depth of about 29,000 feet (5.5 miles or 8.8 kilometers). We know this, because Mount Everest currently stands at 29,029 feet above sea level, and 15 cubits more adds another 22 feet, but as we'll explain below Everest has grown a bit since Noah took to sea. How could this scenario possibly take place without a "magic wand"? Let's see how far we can get.

First, could Noah really be six hundred years old? Senescence is not fully understood, but much of what we know suggests that he could not live to that age in years as we understand them. Normal human cells can only divide between 40 and 60 times. This is known as the Hayflick Limit. At that point the telomere—a kind of doomsday clock on a cell—reaches its end and the cell undergoes apoptosis, or programmed cell death. There are some cells that avoid this fate, but they are not welcome in our bodies. Such cells multiply for their own benefit. We call them cancer.

So, rather than revise normal human biology, let's assume that the Earth was both spinning faster and orbiting the Sun faster—by a factor of about ten. This turns out to be a somewhat useful assumption for what will follow.

Now, we are confronted with a big problem: what would it take to cover every bit of landmass on the Earth, and how might that be achieved?

If we idealize the Earth as a sphere, we can approximate how much water would be needed. First, let's get a measure of the volume of the Earth. (For simplicity's sake, we'll work in metric units. Keep in mind that a kilometer is a little more than half a mile. We could make the calculations even simpler by starting with the Earth's area, but to keep a feel for the process we'll work with volumes.) The tried-and-true formula for finding the volume of a sphere, which we all ~~remember~~ have forgotten from our schooldays, is $(4/3)\pi r^3$. Now, the Earth's radius is 6,371 kilometers. So, plugging numbers into the formula: $[(4 \times 3.14159)/3] \times (6371 \times 6371 \times 6371)]$ equals . . . wait for it . . . okay, that gives us roughly 1,080,000,000,000 (or 1.08 trillion) cubic kilometers. Now, we must figure out how much bigger a sphere would be created if the surface of the Earth were one big ocean overtopping the highest mountains. Here, we run into several tricky points, starting with the highest mountain. Is it really Everest?

That depends. From sea level, yes, but taking into account the bulge of the Earth at its middle, Mount Chimborazo in Ecuador has a claim to being taller. If you were on the moon bouncing a laser beam off the surface of the Earth at various points, the summit of Chimborazo would appear closer than the summit of Everest. If the Earth were spinning faster in those days, the bulge would have been even greater. But let's put aside mighty Chimborazo, because what we want to know is how much extra water it would take to cover all the mountains. Under the centrifugal force of the Earth's rotation, water will bulge just like land, so sea level should be the relevant measure. In those terms, Everest remains the champ, standing 29,029 feet above sea level. However, as I mentioned above, we can't assume that it was precisely that height some six thousand to ten thousand years ago, when the Great Flood presumably happened.

Geologists say that the Himalayan Range, including Everest, was thrust up at least 30 million years ago, when the Eurasian plate collided with the Indian subcontinent plate. In recent years, two opposing forces have been manipulating Everest's height. Lingering upthrust pushes the mountain higher by a few millimeters each year, while erosion caused by the snows that perpetually fall on its peak files it down ever so slightly. Call it a wash.

Even so, in converting from feet to the metric system, we'll round up the height of Everest from 8.85 kilometers to 9 kilometers to account for those extra cubits the Bible mentions and to make sure the peak doesn't pop up during low tide. How much difference do those nine additional kilometers make? We need to run the numbers again. (Skip the formula if numbers make your brain hurt.)

Let's see: [(4 x 3.14159)/3] x (6380 x 6380 x 6380)] equals . . . here it comes now . . . 1.09 trillion cubic kilometers (rounding up a bit). Subtract the 1.08 trillion cubic kilometers of the Earth's interior volume and we have our answer: we'd need roughly 10 billion cubic kilometers of extra water to flood the Earth so as to cover all the mountains.

That's a colossal amount of water. A really, really huge quantity. It's more than all the water on Earth. A lot more. More than seven times as much. At present, the Earth has a bit more than 1.33 billion cubic kilometers of water, with oceans holding roughly 97 percent of that and the remainder found in clouds, glaciers, lakes, rivers, aquifers, etc.

Even though the difference in area between the surface of the Earth and the enveloping sphere of extra water is tiny (about 0.3 percent), there is a significant difference in terms of volume. That's because volume grows much faster than surface area as you scale up—by the cube versus the square. (It's why hummingbirds can fly, and pigs cannot.)

We do get a little break, however. The protruding mountains and landmasses themselves displace some of the volume in question, so we can trim our water budget. Not by much, though. Land above sea level only occupies about a third of the Earth's surface, and of that most lies in plains. Let's be generous, though, and give ourselves a 20 percent discount on the amount of water needed.

So, where are we going to find 8 billion cubic kilometers of water? That figure still represents roughly six times the amount of water known to exist on Earth. The Moon is big enough to hold that quantity, but transferring such a volume (representing well over 10 percent of the satellite's mass) to the Earth would have sent the Moon spiraling away from us. If it had been closer to begin with the tides would have swamped the Holy Land long before Noah laid a single cubit of timber down.

Some have argued that we don't need more water at all. The continents, they say, simply lowered beneath the waterline for 150 days and then popped back up. There are a few problems with this idea. First, rock is denser than water (that's why it doesn't float). Therefore if you were to lower the continents, you would cause the Earth to spin faster, like a spinning skater who pulls in her arms. Then, when you pushed them back up, it would slow again. Since everything on the surface of the Earth shares the rotational momentum of the planet, this would be like a cargo plane taking off and landing with a plastic swimming pool in its hold. To say the water would slosh around is understating it. Then there is the unimaginable displacement of all that water as the continents rise and fall. Think of a fat man doing a cannonball into a child's pool. Noah's Ark would have been swamped by the largest waves ever seen. Surf's up, dude.

The idea of continents sinking and then rising again in a matter of months is a nonstarter, anyway. No natural process could account for such antics. That being so, we're still in the market for water. There is a plentiful source, way out in the Oort Cloud. There, beyond Pluto, in the inky depths of space, lurk innumerable comets, largely composed of water. Comets are like dirty snowballs floating in space. Suppose, then, that two of them were perturbed in such a way that they came streaking toward the Sun and collided with each other just ahead of our planet? Their collision would cause their water content to vaporize and then flash freeze in a beautiful cloud of ice crystals. If the respective momenta were just right, the Earth might overtake the cloud and experience forty days and forty nights of worldwide rain.

There are at least two problems with this idea. First, if we were to dump that much fresh water into the ocean, its salinity would vanish. Ocean life is adapted to living in saltwater; ocean-dwelling fish, marine mammals, plants, and crustaceans can no more live in freshwater than you could in a nitrogen-only atmosphere. Yet, we know that the ocean waters gained their saltiness from interaction with the Earth's minerals. Finding a presalted comet might be like expecting a hen's egg to come with its own seasoning in the shell.

Yet, we have to assume that the rainwater of the Great Flood had the same salinity as the oceans; otherwise, Noah would have had to take aboard all kinds of whales, sharks, and every other kind of marine creature that depends on saltwater for life. A single blue whale would be nearly a fifth of the length of the Ark (30 meters versus 158 meters).

Assuming Noah could install a tank for the pair of them, a third of the Ark's storage capacity would be gone. As for stocking enough krill to feed the blue whales . . . fuggeddaboutit. Noah's got troubles enough keeping the crocodiles from polishing off the pigs. As for the Komodo dragons . . . don't get me started. We have to go with the saline hypothesis.

There may yet be a way to boil up a salt solution. It was long thought that comets delivered the water that today covers two-thirds of the Earth's surface and makes up the clouds, ice caps, and other features of the hydrological cycle. We know the Earth could not have formed with all that water intact. Its early days were hellishly hot and (as the Moon's pockmarked surface attests) under constant bombardment. Nearly all the primordial waters would have been blasted away into space.

Later, when cooler volcanic heads prevailed, water returned to the Earth. But how? It's long been thought that comets delivered it. However, in 2014 a close-up inspection of a comet by the Rosetta Spacecraft threw cold water on that idea. The hitch is this: there are different kinds of water. Most water is made of an oxygen atom plus two ordinary atoms of hydrogen. But a tiny fraction of it is made of oxygen and deuterium, or heavy hydrogen. What makes the deuterium heavy is the addition of a neutron to the proton at the atom's core. The ratio of regular hydrogen to deuterium in the comet water that Rosetta inspected differed a lot from the ratio we find on Earth.

That led scientists to think again. Maybe asteroids rather than comets delivered the water we find on Earth. Asteroids are largely made up of rocks and minerals. So, could they deliver salt along with water? This is at least conceivable. Would we then be able to turn that cloud into forty days' and forty nights' worth of rain, amounting to 10 billion cubic kilometers of water? Improbable, to say the least. Friction is a major problem. That much mass traveling through our atmosphere in such a short period would surely overheat and turn the world into a deadly steam bath. There is a period early in our planet's history when huge volumes of rain fell as the nascent Earth cooled. It rained and it rained and it rained, filling the oceans to the brim. It took a while, though. The deluge evidently went on for thousands of years. Even the most dedicated YMCA fan would have quit the steam bath by then.

Let's suppose that by a *nearly* miraculous coincidence the saltwater crystals left by some deuterium-laden comets required exactly as much heat to melt as they generated by falling through the Earth's atmosphere, and that they went from ice to liquid water in a single step, bypassing the usual vapor path to rain. Let's also suppose that in doing so they came in at an angle that slowed the Earth's orbit and rotation just enough to allow for Noah's age. That would raise still more problems, mentioned above, about friction and rotational momentum, but we'll let them go. We'll assume that it was an almost miraculously gentle braking action— one that gives more time for the rain to fall, as each day becomes longer than the last.

Of course, in saving the ocean's creatures, we raise the necessity of a huge number of onboard aquariums to accommodate all the freshwater fish, otters, turtles, and other river dwellers that would die in a deluge of saltwater. It's not clear how Noah would have built aquariums, but maybe he assigned that task to his sons.

That's only the beginning of the troubles. Boarding is one thing; feeding pairs of every kind of animal on Earth for a little more than a year is quite another. A single elephant, for example, eats up to 600 pounds of fodder a day. For a pair of elephants, Noah would have needed to stow around 10,000 bales of hay on the ark, or risk them going hungry. You really don't want a hungry bull elephant rampaging on your ark.

But let's skip past the onboard agricultural challenges to face up to the biggest of them all: how do we get rid of a volume of water far greater than all the world's oceans in a matter of months? Here's what the Bible tells us happened:

But God remembered Noah and all the wild animals and all the domestic animals that were with him in the ark. And God made a wind blow over the earth, and the waters subsided; the fountains of the deep and the windows of the heavens were closed, the rain from the heavens was restrained, and the waters gradually receded from the earth. At the end of one hundred fifty days the waters had abated; and in the seventh month, on the seventeenth day of the month, the ark came to rest on the mountains of Ararat. The waters continued to abate until the tenth month; in the tenth month, on the first day of the month, the tops of the mountains appeared.

You might think that after all that time on board, with a cacophony of shrieking peacocks, howler monkeys, roaring lions, and trumpeting elephants (not to mention the flies), everybody would be good and ready for some shore leave. But no. According to Noah, it's still too soggy outside. The old hydrophobe makes them all wait another forty days before he even tests for dry land . . . ever so cautiously:

Noah opened the window of the ark that he had made and sent out the raven; and it went to and fro until the waters were dried up from the earth. Then he sent out the dove from him, to see if the waters had subsided from the face of the ground; but the dove found no place to set its foot, and it returned to him to the ark, for the waters were still on the face of the whole earth. So he put out his hand and took it and brought it into the ark with him. He waited another seven days . . .

They're on a mountain, fer chrissakes. Can't they at least get out and stretch their legs?

. . . and again he sent out the dove from the ark; and the dove came back to him in the evening, and there in its beak was a freshly plucked olive leaf; so Noah knew that the waters had subsided from the earth. Then he waited another seven days, and sent out the dove; and it did not return to him any more.

Finally! I lost track of how many days and nights this took, but relying on the good apologists at DefendingGenesis.org, I believe it was no less than 335 days from the end of the flood to day of disembarkation. That's a hellishly long time for anyone to be cooped up on an ark. But is it truly enough time for all that water to recede?

Problem. The mutual attraction law of gravity creates a one-way street. (Or warped spacetime, if you prefer to be Einsteinian, but in this context it amounts to the same thing.) Anything can fall onto the planet, but without the force of a rocket, nothing can get off. To leave the planet, an object has to exceed the minimum escape velocity, which is just over 25,000 miles per hour (or 11 kilometers per second). At that speed, you could travel from New York to Los Angeles in about seven minutes. Don't count on being served a drink along the way.

It takes enormous power to accelerate objects for long enough to get out of the clutches of Earth's gravity. The Saturn V rocket, the most powerful in NASA's fleet, needed to generate nearly 30 kilos of thrust for every kilo of cargo it put into low-Earth orbit. Keep that in mind as we click through some staggering numbers: a cubic meter of water weighs a metric ton. Now, you might think that a cubic kilometer has a thousand cubic meters in it, but that would be wrong. A cubic kilometer runs to a thousand meters per side, but within the box that forms are a million cubic meters. So, to shift 8 billion cubic kilometers off the surface of the Earth, we'd need at least 30 x 1,000,000 x 8,000,000,000 kilos of force. That's 2.4×10^{17}, a number so big that it doesn't have its own name. We call it a hundred quadrillion. It's, uh, ten times a million times a billion. In other words, it's abso-freakin'-lootely huge.

Yet, there are forces in the universe capable of the job. The solar wind, for example, has blown most of the atmosphere and water off the surface of Mars. Of course, that's taken billions of years, and the gravity of Mars was only about 40 percent of Earth's to begin with.

There is at least one cosmic tool that can get the job done on time. A gamma ray burst could flash-vaporize and blow the water away in a jiffy. Gamma ray bursts are somewhat mysterious cosmic lightning bolts, carried by high-energy photons streaking across space. They are thought to emanate from stars with nearby massive planets that wind up their magnetic fields until they snap.

The trouble is that a gamma ray burst with the power to blow away that much water would sterilize the planet like God's own autoclave. The laws of thermodynamics say that you just can't heat up water that fast and blow it away without the heat spreading. The energies required for the job would be globally catastrophic. Mount Ararat, where Noah's Ark supposedly came to rest, would be the scene of a flash-fry barbeque of every living creature left on Earth. Pass the Famous Dave's . . .

Could the water have drained into vast caverns under the sea? That idea has been kicked around, but it's a no-hoper. The Earth's gravity draws everything toward its center of mass. Only the interposition of something denser prevents an object from continuing its journey to the center of the Earth. You've probably known this since childhood days in the bath. If you press a rubber ducky to the bottom, it will pop up when you release it, because the air inside it is much less dense than the water in the bath.

It's true that electromagnetic forces, being far stronger than gravity, can temporarily offset gravity's pull for some objects. A steel truss bridge, held together by electromagnetic force between its atoms, can keep a roadway suspended above the air for a lifetime or more. But even the best bridge has a load limit. Any attempt to hide a vast ocean of boiling water (it's hot down there!) under a crust of rock would eventually give way. The first volcanic eruption would spurt all that water back up to the surface as the rock pressed downward, and it would be steam bath time all over again.

In any case, we know that there isn't a second ocean hidden under the Earth's crust. Even though no one has ever drilled into our planet's interior (the crust is just too thick), scientists have mapped it in detail.

How? The same way that an expectant mother can get an image of her fetus: by sonogram.

When earthquakes occur, seismic waves rattle through the Earth, and seismologists can read the information they carry to the surface to create a portrait of the Earth's core.

The only other natural solution would be a black-hole vacuum cleaner. Black holes can, in principle, absorb a limitless amount of stuff. (In practice, astrodynamics limits them to gobbling up a decent chunk of a galaxy before the stellar winds and rotational forces put other stars out of reach.) So, in principle, a black hole could soak up all the excess water and leave Noah high and dry. The trouble is that, unlike a Hoover, black holes don't have an off switch. Any black hole that started to feast on the Earth's bounty just wouldn't quit. It would consume the entire planet like a glutton slurping up a plate of spaghetti. Black holes are a dead end.

But here's the weirdest thing of all. In a passage that never seems to get quoted, the Bible tells us that at the end of the voyage, Noah roasted at least one of every kind of animal:

> Then Noah built an altar to the Lord, and took of every clean animal and of every clean bird, and offered burnt offerings on the altar. And when the Lord smelled the pleasing odor, the Lord said in his heart, "I will never again curse the ground because of humankind ... " (Genesis 8:20–21, NRSV)

This makes it highly improbable that we would have any elephants, pandas, polar bears, or orangutans today. These slow-reproducing creatures are quite unlikely to have become impregnated or given birth during the voyage—especially not with all the other animals watching.

So, in the end, we are stumped. What a wild and pointless ride. After all that, the world is still full of sin and cockroaches. No wonder Noah got blind drunk when it was all over.

The sober truth is this: either Noah's Flood is a tall tale gone wild, or God is a magician after all. But let's not quit now. There's so much more fun in Bible stories yet to come. And in the end, we shall find it may be that with Nature, all possible things are.

3

 Sodom and Gomorrah

The story of Lot, that good man of Sodom, has come to be the iconic morality tale that nobody actually reads. Lot's hometown is a byword for orgies. Indeed, British common law employed the term "sodomy" to denote male gay sex. But taken afresh, Genesis 19 offers up a story that is exceedingly strange, in both morality and science.

It begins with a pair of angels visiting Lot. What exactly are angels? There seems to be consensus that angels are messengers from God, spiritual beings without corporeal form. This being the Old Testament, we turn to Rabbi Baruch S. Davidson for more:

> The Jewish belief in angels goes as far back as the Book of Genesis. . . . According to Jewish tradition, an angel is a spiritual being and does not have any physical characteristics. The angelic descriptions provided by the prophets—such as wings, arms etc.—are anthropomorphic, referring to their spiritual abilities and tasks.

Christmas ornaments notwithstanding, Christian theology appears to concur. This raises problems for the story. Lot sees the angels approaching the gates of Sodom and bows down in humble greeting. And with that, we got trouble, right here in sin city.

It begins with the old Casper-the-Ghost problem: if angels have no physical form, how do they make themselves seen and heard? We see objects because light bounces off them, carrying away information about each object, such as its shape, texture, and color. Voices can be heard only because air is forced by our lungs through a vibrating larynx and then modified by tongue and teeth. It's true that a diode can emit light and a speaker can vibrate much like a larynx, but without any physical form this just can't happen. There's no "there" there.

If you suppose that angels are made of spiritual "stuff," and that this is just different from material stuff, you run into the interaction problem. The universe, we now know with reasonable certainty, is composed of a whole lot of stuff that balances out exactly. Physicists Alexei V. Filippenko and Jay M. Pasachoff, authors of *The Cosmos: Astronomy in the New Millennium*, explain it this way:

> All of [the matter and energy] particles consist of positive energy. This energy, however, is exactly balanced by the negative gravitational energy of everything pulling on everything else. In other words, the total energy of the universe is zero! It is remarkable that the universe consists of essentially nothing, but (fortunately for us) in positive and negative parts. You can easily see that gravity is associated with negative energy: If you drop a ball from rest (defined to be a state of zero energy), it gains energy of motion (kinetic energy) as it falls. But this gain is exactly balanced by [a rising] negative gravitational energy as it comes closer to Earth's center, so the sum of the two energies remains zero.

We don't understand all the details—gravity appears to work in reverse in "empty" space, where virtual particles predominate—but one thing seems to be incontestable: energy is conserved. Or, as you may have heard while dozing off in high school physics, energy can neither be created nor destroyed; it can only change forms (one of those forms being matter, which is essentially frozen energy).

But if you introduce spiritual stuff into the universe, suddenly you're breaking that law. When angels speak, or flap their putative wings, or merely deflect photons from their preordained path, they violate the conservation laws. That can only mean one thing: it's a miracle! And possibly a disaster as well. Physicists surmise that our universe got its start because of a slight deviation from perfect equilibrium, and you know what happened then—the Big Bang! A disruption of the balance between positive and negative energy could prove far more disastrous than the fate that awaits Sodom and Gomorrah.

Of course, Rabbi Davidson makes no such claim about spiritual energy. He says angels have no "physical characteristics"—in which case they cannot possibly have physical consequences. So, for now let's give Lot the burden of belief. Let's suppose that he only sees these angels in his mind's eye. Back to the story:

> Behold now, my lords, turn in, I pray you, into your servant's house, and tarry all night, and wash your feet, and ye shall rise up early, and go on your ways. And they said, Nay; but we will abide in the street all night. And he pressed upon them greatly; and they turned in unto him, and entered into his house; and he made them a feast, and did bake unleavened bread, and they did eat. (Genesis 19:1–4, KJV)

Well, that didn't last long. It is impossible to explain how incorporeal angels could eat a meal. Impossible, that is, for anyone constrained by reason and the findings of science. We might suppose that by some inexplicable means they can morph into human form temporarily, but then we have left science far behind. So, we'd best assume they are angels in name only—that is, humans tasked with delivering a divine message.

At any rate, to the townspeople they appear to be human—or at least fresh meat. Our story continues:

> But before they lay down, the men of the city, even the men of Sodom, compassed the house round, both old and young, all the people from every quarter: and they called unto Lot, and said unto him, Where are the men which came in to thee this night? Bring them out unto us, that we may know them. (Genesis 19:4–5, KJV)

Well, you *know* what "know" means in the Bible! Our angels are in for some Sodomite hospitality. But, ever the quick thinker, Lot comes up with a plan to deter the would-be angel-raping mob. Now, remember, Lot is the one really good man in town, so his actions, we must presume, are exemplary. Here's what he does:

Lot went out at the door unto them, and shut the door after him, and said, I pray you, brethren, do not so wickedly. Behold now, I have two daughters which have not known man; let me, I pray you, bring them out unto you, and do ye to them as is good in your eyes: only unto these men do nothing; for therefore came they under the shadow of my roof. (Genesis 19:6–8, KJV)

The moral message is clear if stunning: such is the law of Sodomite hospitality that if you have guests staying in your house and they are threatened with gang rape, better you should throw your young, virginal daughters to the mob than allow your guests to come to harm. There are two odd things about this, other than the shocking idea that involuntary victim-substitution is a morally sound practice. First, turns out these "virgins" are already married. It can't be that Lot is lying, for he is a good man, and we know that the commandments include, "Thou Shalt Not Lie." Second, it never occurs to him that this mob, seemingly bent on homosexual rape, might be more interested in his sons-in-law than in his daughters.

At any rate, Lot's offer is vehemently rejected, and it soon appears that the good man himself will be added to the list of victims. The leader of the mob cries, *"Now will we deal worse with thee, than with them. And they pressed sore upon the man, even Lot, and came near to break the door."*

Things aren't looking good for our hero, but fortunately for him his guests are no patsies: *"the men put forth their hand, and pulled Lot into the house to them, and shut to the door. And they smote the men that were at the door of the house with blindness, both small and great: so that they wearied themselves to find the door."*

Could the angels have blinded the mob without magic? Perhaps. A little flash powder might have proved useful, but that wouldn't be invented until long after the Bible was written. Still, if one of them flung a sack of fine ash from a window, it might have loosed a blinding, stinging cloud that enveloped the crowd.

Apparently the mob disperses, because the next thing we know, the angels are advising Lot to go back outside—along with his sons-in-law! That's how we come to know that the daughters are married.

And the men said unto Lot, Hast thou here any besides? Son in law, and thy sons, and thy daughters, and whatsoever thou hast in the city, bring *them* out of this place: for we will destroy this place, because the cry of them is waxen great before the face of the Lord; and the Lord hath sent us to destroy it. And Lot went out, and spake unto his sons in law, which married his daughters, and said, Up, get you out of this place; for the Lord will destroy this city. But he seemed as one that mocked unto his sons in law. (Genesis 19:12–14, KJV)

In other words, these layabouts think he's kidding. They're not about to lose sleep over an old man's ravings. Come dawn, the angels have had enough. They shove Lot, his wife, and two daughters out of their home. The sons-in-law? Not so much. In what is perhaps the earliest demonstration of the saying, "Ya snooze, ya lose," the young men are left behind to a fate possibly worse than gang rape. Ah, the Bible. It's comedy gold.

Lot has finally been persuaded that Sodom is no place to raise a family, but when it comes to the choice of refuge, he's strictly an urbanite. No mountain retreats for him. However, he does know the difference between big city sin and small town virtue, as quickly becomes apparent. Relaying a warning from God, the angels say, *"Escape for thy life; look not behind thee, neither stay thou in all the plain; escape to the mountain, lest thou be consumed."*

But Lot still doesn't want to go to the mountain, so he addresses a plea to God, via his messengers:

Oh, not so, my Lord: behold now, thy servant hath found grace in thy sight, and thou hast magnified thy mercy, which thou hast shewed unto me in saving my life; and I cannot escape to the mountain, lest some evil take me, and I die: behold now, this city *is* near to flee unto, and it *is* a little one: Oh, let me escape thither, *(is* it not a little one?) and my soul shall live.

The Lord decides to cut him a break:

See, I have accepted thee concerning this thing also, that I will not overthrow this city, for the which thou hast spoken. Haste thee, escape thither; for I cannot do any thing till thou be come thither. Therefore the name of the city was called Zoar. (Genesis 19:18–22, KJV)

Off they go, Lot, his wife, and his daughters, on a one-way trip to Zoar. You might think that the daughters, who will never see their husbands again, would be the ones tempted to look back, but no, it's his wife. Why her? Not for us to stir scandal.

Famously, for the sin of looking back, Mrs. Lot gets turned into a pillar of salt. Like so many other women in the Bible, Mrs. L goes unnamed. However, in the Midrash, a body of Jewish commentary on the Torah, she is called Edith. (Perhaps her husband's first name was Archie.) There has also been much discussion down the centuries about just what this turning-into-salt business could mean. A pillar, some commentators believe, might have meant a statue, or even her body rendered in salt. But others believe she was turned into a literal pillar of salt. Keep that in mind as we advance into the inferno.

The sun was risen upon the earth when Lot entered into Zoar. Then the Lord rained upon Sodom and upon Gomorrah brimstone and fire from the Lord out of heaven; and he overthrew those cities, and all the plain, and all the inhabitants of the cities, and that which grew upon the ground. (Genesis 19:23–25, KJV)

There are three good scientific candidates for explaining how a wroth Yahweh waxed the Cities of the Plain: volcanic eruption, meteor impact, or a mighty storm.

Surprisingly, the Middle East has more than its share of volcanoes, though eruptions are certainly not an everyday occurrence. As weather is an everyday experience, let's go with the storm first.

Biblical scholars have long recognized that "fire" in the Bible can mean lightning, since the ancients had no notion of electricity or how it differs from flame. Indeed, since a lightning strike often causes fire to break out, they had good reason to associate the two. The phrase "rained upon Sodom and upon Gomorrah" certainly adds to the impression of a storm. But what the heck's brimstone? It's an antique name for sulfur, an element with the color of yolk whose odoriferous compounds smell of rotten eggs. Sulfur happens to be commonly found in crude oil—and little wonder, since oil is a byproduct of organic decomposition—and organisms, such as eggs, contain sulfur.

Jean LeClerc (1657–1736), a Swiss theologian, first thought of the connection. He argued that lightning might have struck bituminous soil in the vicinity of Sodom and touched off a stinking inferno. There is little coal in the region, but oil? That's another matter. Oil is the runner-up liquid of the Middle East, coming a close second to spilled blood. And when oil dries out, what you have left is bitumen, a semisolid blob of flammable gunk.

There is no scholarly consensus on the location of Sodom and Gomorrah, but various references in the Bible suggest that they were in the vicinity of the Dead Sea. There's even a tall hill on the south side known as Mount Sodom, and not far from that are bituminous deposits. Better yet, Mount Sodom's full of salt!

Could a mighty storm have struck the Cities of the Plain? Certainly. It's true that rain is rare in the area: that's why the sea is dead. But there could have been no cities if it didn't rain sometimes, and even now there are a couple of downpours annually. That they are rare makes the storm scenario more compelling.

Could a stray lightning bolt have touched off a pocket of natural gas and bitumen?

Quite conceivably. Could another bolt of lightning have struck Lot's wife? Hey, why not? Especially if she were carrying Lot's golf clubs.

In the contemporary United States, the actuarial chances of being hit by lightning at some point during a lifetime are 1 in 3,000. But the chances go way up if you're swinging a mashie niblick on the links . . . or just standing upright on a plain during a storm, casting a longing gaze at the home you have just abandoned.

Lightning is complicated. It starts with "feelers" that make their way to the ground. Once they touch down, a huge backflow of electrons from positive ions rushes up from the ground—often taking a more efficient route than the feeler. Someone standing on level ground makes a convenient jumping off point for the upward bound jolt. It's even better if they are holding a golf club, but it's more plausible that Edith, knowing that she would never return to good old Sodom, had donned all her jewelry. Middle Eastern jewelry for a middle-class woman would have been set in copper, making Mrs. L a first-rate electrical conductor. And this, in turn, might explain why she turned back. Perhaps she felt a tingly sensation that was a sign of worse to come.

At any rate, a lightning strike does terrible things. More often than not the victim survives, but with burns, burst blood vessels, memory loss, heart arrhythmia, scars, and other nasty effects. But however powerful the bolt no one turns into salt.

There just isn't enough salt in a human body to make a garden gnome, let alone a full-size statue. Even with our high-sodium diets, the salt content of our bodies is around 0.4 percent—about the same as seawater. Extracted, a body's salt would come to just over three-quarters of a cup, hardly enough to make any sort of respectable pillar. Maybe we need to consider salt in a larger context.

In daily use, salt means sodium chloride, but in chemistry there are many kinds of salt. They form when ions with too many electrons meet up with ions with too few and form a crystalline union. (There are also liquid salts, but we'll ignore them, since you can't make a liquid statue.) Now, intriguingly, fossils are sometimes formed by the replacement of bones and teeth by crystalline minerals, some of which could be salts. However, this process requires a slow drip of water over millennia under ideal conditions—and if one thing's clear in Genesis 19, it's that God is in a sweat to destroy the Cities of the Plain. So we really can't figure Mrs. Lot for a fossil.

A meteor strike certainly could have taken out Sodom and Gomorrah. When large impactors hit the Earth's atmosphere, it's normal for them to break up into smaller chunks, so there's no difficulty in accounting for both cities being leveled. There's some intriguing evidence for this notion. Alan Bond and Mark Hempsell of Bristol University have recently simulated the trajectory of a large meteor described in an ancient Sumerian tablet. They found that it would have streaked low over the Levant, skipped across the Mediterranean, and slammed into a mountain in Austria—which does indeed bear the hallmarks of a low-trajectory impactor. What's more, they say, the plume from a kilometer-sized meteor could have dropped fire and brimstone onto the valley where the ill-fated cities lay. It's plausible but by no means uncontested. However, unless we assume that a meteor chunk hit Mount Sodom, releasing a spray of salt crystals that studded Lot's wife, this tale does nothing to help us explain her fate.

What would it take to transmute Dame Edith into salt? Alchemists searched for centuries for a way to turn base metal into gold, but they were barking up the wrong beaker. It's not a matter of chemistry but of nuclear physics. One element can indeed turn into another. There are two distinct ways for this to happen: from the bottom up, or the top down.

First, you can build up from simple elements—hydrogen is the simplest—to make anything you want. It's a simple process: all you need is a decent-size star. Your run-of-the-mill sun won't do; even at 300,000 times the size of the Earth, it's only capable of fusing hydrogen into helium, along with a few stray elements. Boring . . .

What you need is an M-type red giant. Wait a few million years, then as its supply of hydrogen runs out, stand by for the collapse of its core. Boom! The resulting supernova will shower you with all the elements you could want, including sodium and chlorine, the ingredients of table salt.

Let's assume that the moment the lightning hits, a warp in the space-time fabric slingshots Edith into a red giant just as it is set to detonate. A supernova happens about once every 50 years in our galaxy, but thanks to time dilation and the 100,000-light-year diameter of the Milky Way, we might have our choice of several. Let's also assume that the warp stands by at a respectful distance to return the transformed atoms to the spot where lightning struck.

Red giants can certainly produce sodium and chlorine—in fact, they are the foundries for all of the elements except hydrogen, helium, and lithium. However, there's a catch. Assuming we could get Mrs. L in place at just the right moment for Big Red to go supernova, how could we tell which of the emerging atoms were hers? We all may be unique individuals, but our constituent atoms are not. Nuclear particles are indistinguishable, so when they get mashed into new elements there is no hope that we can pick out the ones that used to be Lot's good lady and build us a pillar of salt. Better try the other method.

The second and far easier way is to wait for spontaneous radioactive decay. Large atoms have an uneasy hive of protons and neutrons at their core. Protons in particular, being of the same charge, all want to get away from each other. But their positive electrical charge is overwhelmed by the strong nuclear force, which binds both protons and neutrons together at very short distances. That works just fine . . . until the weak nuclear force makes mischief. Once in a while—sometimes a very great while—the weak force upsets the balance and an atom splits. Usually, it is a very large and cumbersome atom, say uranium-235, which falls apart. When it does, it releases a flash of energy and debris, leaving behind some daughter elements, which undergo more changes and eventually settle down to a boring, stable career as, say, lead.

It's a long, tedious process. Just to get halfway through a brick of uranium-235, the stuff of the first atom bomb, would take 700 million years. However, as J. Robert Oppenheimer and his fellow Manhattan Project scientists discovered, if you confine the debris—especially the escaping neutrons—into a small enough area, you can cut the required time down to a few milliseconds.

In principle, neutron bombardment can overcome the strong force and break apart any atom with a complex nucleus. So, perhaps by selective use of energetic neutrons it might be possible to retool atoms like carbon . . . but then, instead of becoming a pillar of salt, poor Mrs. Lot might well have turned into a mushroom cloud and cleared the whole valley. Better leave this radioactive tomfoolery to one side.

If we want to end up with a pillar of salt, we can perhaps rely on a volcano. There are more than a dozen hot spots in Syria and Iran alone. One of these, an unnamed volcano in Syria that bears the number 230040, last erupted in 2670 BCE. Is that a plausible date for the demise of Sodom, Gomorrah, and Lot's wife? Could be.

Genesis and the other books of the Pentateuch are very old. Biblical scholar Peter Enns, having revisited all of the work on the subject and then added his own, concludes that the earliest books of the Bible were assembled from both oral traditions and written records by the Jews who survived the exile to Babylon around 586 BCE:

> The exile was arguably the most traumatic and therefore most
> influential historical event in Israel's ancient history. The Israelites

understood themselves to be God's chosen people: they were promised the perpetual possession of the land, the glorious temple as a house of worship, and a descendent of David sitting perpetually on the throne (2 Sam 7:4–16). With the exile, all of this came to a sudden and devastating end. Exile in Babylon was not an inconvenience. It meant to the Israelites that their relationship with God had been disrupted. God could no longer be worshiped as he himself required—in the Jerusalem temple. Israel's connection with God was severed: no land, no temple, no sacrifices, no king. . . . Since these heretofore ties to Yahweh were no longer available to them, the Israelites turned to the next best thing: bringing the glorious past into their miserable present by means of an official collection of writings. Some of these writings were collected or edited during the exile or afterward, while others were composed during those times. But the trauma of the exile was the driving factor in the creation of what has come to be known as "the Bible."

In short, the story of what happened to the Cities of the Plain might have been around for a few millennia, but the version we know as canonical didn't take shape until the middle of the first millennium BCE. Could a volcano underlie the whole story? These geological dragons erupt in many different ways. Some, like Hawaii's Kilauea, are pretty tame. It burps and steams, but its lava flow is so slow that people can simply walk away. There's even a national park there for lava gawkers.

Others, like Washington State's Mount Saint Helens, go off like the aforementioned nuclear bomb. When these so-called Plinian eruptions occur, they can devastate whole cities (I'm talking about you, Sodom and Gomorrah).

There are several ways the volcano can do its mischief. Volcanic bombs—hot chunks of lava—exit the magma conduit like shells from a Howitzer and come raining down on the surrounding landscape. They cool in the air and land as solid rock, crushing people and sometimes whole structures beneath them. It is conceivable that Lot's wife was struck by a sizable volcanic bomb that obliterated her and left a lump of lava in her stead. However, there's no reason to think it would be mistaken for a pillar, let alone one made of salt. Small lava bombs can be football shaped, but the larger ones are generally wider than they are tall on impact. If the Bible told us that Mrs. L had been turned into an anvil, we might suspect a volcanic bomb, but a pillar of salt? Let's keep looking.

Another terrifying way volcanoes can kill is via a pyroclastic flow. This is a tsunami of ash, flying rock, and superheated gas that races down the side of an erupting volcano and devastates everything in its way. It can swoop through a valley at speeds of over 400 miles per hour. If this is what God had in mind, no wonder he advised Lot to scurry up into the mountains.

We know that a pyroclastic flow can mummify any living thing in its path. The ancient Roman city of Pompeii stands as proof of concept. In the year 79 CE, Mount Vesuvius blew its top, sending wave after wave of pyroclastics racing away. The fourth of these turned Pompeii, six miles distant, into a pizza oven. The temperature shot up to 570 degrees Fahrenheit, instantly killing everyone in town. Later waves coated them in ash, which hardened to give us the eerie stop-action remains that make the town famous.

Like an avalanche, a pyroclastic flow tends to cut a well-defined swath. If Mrs. L lagged well behind the rest of the Lot party, it might be that she became engulfed in a pyroclastic flow while the others trod on. It is inconceivable that she could have remained standing upright against the force of the flow, but if she came to rest against a boulder or tree, then maybe, just maybe, she was turned into an upright, ash-clad mummy. Now, a mummy is a lot like a statue, and a statue is a plausible interpretation of "pillar."

And here's a bonus: According to the U.S. Geological Survey, volcanic ash often contains salts of various kinds, including table salt. What's more, salt tends to leach, meaning it would be found on the surface of, say, a mummified mother of two from Sodom.

With that, all the pieces fit together. Sodom, Gomorrah, and the other iniquitous cities of the plain were first bombarded by fiery volcanic debris then demolished by a pyroclastic flow, the outer edge of which turned Edith into the legendary pillar of salt.

And so ends the tale of Lot's wife.

* * *

Epilogue: Having never looked back, Lot eventually goes to the mountains after all. There, that good man of Sodom has sex with his daughters, impregnating them both. The Bible claims it was their idea, but the passage reads like a boozy defense lawyer's brief. Either way, it's hard to imagine making a good Sunday School lesson out of the story of Lot and his family.

And come to think of it, if Lot and his daughters didn't look back—and they couldn't have, or the whole episode makes no sense—how could anyone have known afterward that Edith was turned into a pillar of salt? How do we know that she didn't just grab the opportunity to flee from that perverted lunatic she was shacked up with? Did Lot identify the remains? If so, why did anyone believe him? Sane people don't walk up to geological features and say, "That's no pillar; that's my wife!" The mystery remains.

But we'll always have Sodom.

4

 River of Blood

If I were King of the Forest, Not queen, not duke, not prince.
My regal robes of the forest would be satin, not cotton, not chintz.
I'd command each thing, be it fish or fowl.
With a woof and a woof and a royal growl

—Cowardly Lion, "The Wizard of Oz"

If you were not just king, but the King of Kings, and you wanted to impress that fact on your people by conjuring up a miracle, what would you do? Royalty aims to impress. One thing you would surely not do is perform an act that nature regularly serves up. For example, a lightning bolt striking a heathen holding a staff really should not count as divine wrath. It happens to golfers all the time.

But even more than that, you'd want to avoid miracles that fraudsters have already pulled off. In Exodus, the famous "Let My People Go" chapter of the Bible, God backs Moses in his suit before the Pharaoh by supplying him with a chain of miraculous plagues intended to convince the ruler of Egypt to free his Hebrew slaves. Oddly, though, he warms up with a conjuring trick that is immediately duplicated by the court magicians:

> The Lord said to Moses and Aaron, "When Pharaoh says to you, 'Perform a wonder,' then you shall say to Aaron, 'Take your staff and throw it down before Pharaoh, and it will become a snake.'" So Moses

53

and Aaron went to Pharaoh and did as the Lord had commanded; Aaron threw down his staff before Pharaoh and his officials, and it became a snake. Then Pharaoh summoned the wise men and the sorcerers; and they also, the magicians of Egypt, did the same by their secret arts. Each one threw down his staff, and they became snakes; but Aaron's staff swallowed up theirs. Still Pharaoh's heart was hardened, and he would not listen to them, as the Lord had said. (Exodus 7:8–13, NRSV)

Score: Pharaoh 1, Moses 0. Obviously, the situation now calls for a clear, unambiguous, laws-of-nature-defying miracle. So, it's a bit of a puzzle why the Almighty, in coaching Moses how to bring around a skeptical Pharaoh, offers this plan:

In this thou shalt know that I am the Lord: behold, I will smite with the rod that is in mine hand upon the waters which are in the river, and they shall be turned to blood. And the fish that is in the river shall die, and the river shall stink; and the Egyptians shall loathe to drink of the water of the river. And the Lord spake unto Moses, Say unto Aaron, Take thy rod, and stretch out thine hand upon the waters of Egypt, upon their streams, upon their rivers, and upon their ponds, and upon all their pools of water, that they may become blood; and that there may be blood throughout all the land of Egypt, both in vessels of wood, and in vessels of stone. And Moses and Aaron did so, as the Lord commanded; and he lifted up the rod, and smote the waters that were in the river, in the sight of Pharaoh, and in the sight of his servants; and all the waters that were in the river were turned to blood. And the fish that was in the river died; and the river stank, and the Egyptians could not drink of the water of the river. (Exodus 7:17–21, KJV)

Okay, major inconvenience, certainly more disruptive than, say, falling frogs (we'll come to that), but is this a miracle? Nope. It happens all the time. The river of blood "miracle" remains a mystery, not because we can't explain it but because we so easily can. Rivers run blood red with remarkable frequency. Now, granted, blood and blood red are two different things, but how could anyone in Pharaoh's time tell the difference? There were no microscopes or chemical analysis kits to deploy. Seeing a river turn red and all the fish die would be evidence enough for the ancients.

In 2014 China saw its third such incident in as many years, according to reporter Katie Sola:

> Locals in the Chinese city of Wenzhou woke up to a mysteriously blood-red river Thursday morning. According to China Radio International, the river began to redden at 6 a.m. local time, baffling residents who have spent their entire lives in the town of Wenzhou in east China's Zhejiang Province. Locals complained that they could no longer fish in the river . . .

That particular incident appears to have been a case of industrial dumping, but there are at least two natural causes of such events as well. One happens when a river's cut bank, the steep outside bank of a curving water channel, collapses. If the soil above the waterline happens to be rich in red clay, hey presto! You get a blood-red river.

In North America, we have not one but two Red Rivers, so named because of the clay-filled soils that imbue them with a reddish hue when they flood. Not only that, but the southern Red River is bilingual. To quote from the Texas Historical Society, theirs "is the second longest river associated with Texas. Its name comes from its color, which in turn comes from the fact that the river carries large quantities of red soil in flood periods. . . . The Spanish called the stream Río Rojo, among other names."

However, given the specifics of the Bible story, the most plausible explanation is an algal bloom. You've probably heard of this phenomenon as a "red tide" that frequently afflicts the Gulf of Mexico.

Algae are microscopic organisms that live in waters ranging from oceans and lakes to rivers and untended swimming pools. Like us, they are eukaryotic, meaning their cells have nuclei, and like us they are spectacularly diverse. Actually, they are way more diverse. As much as you might think Rush Limbaugh differs from Kanye West, next to algae those guys look like identical twins.

Algae range from unicellular creatures, including the crystalline diatoms, as varied as snowflakes, to the giant kelp, sentinels of the underwater forests. Some varieties of algae known as dinoflagellates have a nasty habit: when the weather turns hot, they create so-called red tides. The waters literally turn red as the tiny cells multiply exponentially. Some species are harmless, but others foul the waters—their density increases, they signal each other chemically and then on cue begin to poison the waters. Science educator Monica Bruckner of Montana State University notes:

> Red tide algae make potent natural toxins. It is unknown why these toxins are created, but some can be hazardous to larger organisms through the processes of biomagnification and bioaccumulation. Grazers such as fish and krill are unaffected by the toxins, so as they eat the algae the toxins are concentrated and accumulate to a level that is poisonous . . . to organisms that feed on them. Large fish kills and several mammalian diseases and deaths have been attributed to consumption of shellfish during red tide algal blooms.

Harmful algal blooms, the term scientists prefer, have become a regular summer event in the Gulf of Mexico. Here's an excerpt from a 2014 Associated Press account:

> A massive toxic algae called red tide is killing sea turtles, sharks and fish in the northeast Gulf of Mexico and is threatening the waters and beaches that fuel Florida's economy. Fishermen who make a living off Florida's coast are reporting fish kills and reddish water.
>
> This particular strain of red tide, called Karenia brevis, is present nearly every year off Florida. . . . Red tide kills fish, manatees and other marine life by releasing a toxin that paralyzes their central nervous system. The algae also foul beaches and can be harmful to people who inhale the algae's toxins when winds blow onshore or by crashing waves, particularly those with asthma and other respiratory ailments.

Harmful algal blooms aren't confined to the oceans. They occur in rivers, lagoons, and other waterways as well. The Nile, on which Egyptian civilization was built, was especially hospitable to algae looking to hold a rave, because until the Aswan Dam was built it flooded annually. During the flood season, the Nile became turbocharged with nutrient rich soils washed into the river by torrential rains upstream. At such times, the Nile would actually take on a reddish hue just from its nutrient load. For algae, warm weather and a brimming river can only mean one thing: it's party time!

Could that be what happened when Aaron waggled his rod over the waters? Colin Humphreys, author of *The Miracles of Exodus*, certainly thinks so.

> The first plague occurred not on the freshwater Nile River, but in the Nile Estuary exactly where toxic algae are able to live. The time of year . . . was probably July to September, the hottest part of the Egyptian summer, during the Nile inundation when nutrient rich soil was brought down the Nile. The most likely time was September, when the Nile flooded and the nutrients would have been at a maximum. So I suggest that at the time of the first plague, toxic red algae were caused to bloom by this combination of very hot weather and the nutrient-rich water. The resulting toxins killed the fish, just as in the estuaries of the U.S. more recently, and the dead fish caused the river to stink just as described in the book of Exodus.

There is one other reason to favor the algal bloom over a literal transformation of water into blood. Real blood clots. If you're looking for a miracle, coagulation is a pretty good candidate. Or, it would be if we didn't understand it so thoroughly.

Blood flows freely through veins and arteries, and to the unaided eye it looks like a liquid through and through. In reality, however, blood is an Irish stew of ingredients. In addition to red and white blood cells, blood teems with platelets, fibrinogens, and dozens of protein factors all just waiting for a blood vessel rupture so they can go into action. If you cut your finger while dicing tomatoes, for example, then unless you happen to be descended from European royalty, you'll notice that right away the coagulation cascade begins. By the time you've run out of curses and stopped dancing a jig around the kitchen, a clot may already have formed.

If you trace your lineage to the inbred courts of Europe, you may find that you're a hemophiliac in need of an occasional clotting-factor infusion. For most people, though, along with all other mammals and many other animals, coagulation is a naturally occurring part of their body's functioning. The cascade involves a series of biochemical axes, called proteases, cutting through protein safety cords that prevent blood from clotting in your veins. Once the cascade begins, several interesting things happen. Platelets unfurl to resemble ninja stars that pile up on each other at the wound site. Fibrinogens turn into stringy, sticky fibrins that weave themselves around the platelets and hold them fast in place. Sticky white blood cells also pile on.

What triggers clotting? Apart from the body's internal injury signals, a major prompt is of course exposure to air. And in Egypt, the Nile fans out to provide maximal exposure to the atmosphere. Had the rod of Aaron turned it to real blood, all of Egypt would have been one gigantic scab. Now that would have been a miracle! Given that 300 million cubic meters flow down the Nile every day—as much as the daily groundwater use in all of the United States in 1990—that scab would have made one hell of a tourist attraction. People would be clambering over it to this day, while the pyramids sulked in silence.

So, an algal bloom remains the best bet to explain the Bible's river of blood. Of course, all this reflects modern scientific knowledge. Ancient Egyptians wouldn't have known *Karenia brevis* from *Anna Karenina*, let alone dreamed of the existence of microorganisms. But they did know the Nile. Algal blooms have been rouging the mighty river for thousands of years. Somebody claiming to represent a god more powerful than Pharaoh who attempted to prove it by turning the Nile to "blood" would have been about as impressive as a rainmaker in Seattle who produces an afternoon shower.

Indeed, the very next line of the Bible tells us: "But the magicians of Egypt did the same by their secret arts; so Pharaoh's heart remained hardened, and he would not listen to them, as the Lord had said." Poor Aaron. After all that effort to impress Egypt's ruler, he must have been ready to snap his rod in two and toss it into the Nile.

5

 What's the Buzz?

In our last chapter the Lord tried spooking Pharaoh into letting the Hebrews go, but seeing a river go red was never going to change the pharaonic mind. Stinking fish or no stinking fish, the answer is uh-uh.

There's no mystery about the Pharaoh's hard-heartedness. God is stage-managing it:

> And the Lord said unto Moses, See, I have made thee a god to Pharaoh: and Aaron thy brother shall be thy prophet. Thou shalt speak all that I command thee: and Aaron thy brother shall speak unto Pharaoh, that he send the children of Israel out of his land. And I will harden Pharaoh's heart, and multiply my signs and my wonders in the land of Egypt. (Exodus 7:1–3, KJV)

Here is the real puzzle. If the fouling of the waters didn't work, then presumably the Lord could have popped up in the Pharaoh's court, worked an unmistakable miracle, and helped the Hebrews hit the road. For example, being all knowing, God must have been cognizant that one day people would invent television. If he really wanted to convince Pharaoh once and for all, he could have had Aaron waggle his rod and hey, presto! There's Bill O'Reilly on a 50" flat-screen ranting about the "War on Christmas." Now that would have scared Pharaoh straight.

Could something as complex as a flat-screen TV, complete with a Fox News broadcast, materialize out of thin air? Had we but world enough, and time, as Andrew Marvell might have put it, no problemo.

Physicists have calculated that given particles and time enough, random fluctuations in the universe will eventually generate whole human brains, complete with thoughts and memories. If the universe can do that, then surely God can whip up a big-screen talking head on command. That's technomancy!

But no. God makes it clear that they and the Egyptians will have to suffer through the whole show, 'cause he's got something to prove. Poor Pharaoh has no choice in the matter; it's all foreordained:

> When Pharaoh does not listen to you, I will lay my hand upon Egypt and bring my people the Israelites, company by company, out of the land of Egypt by great acts of judgment. The Egyptians shall know that I am the Lord, when I stretch out my hand against Egypt and bring the Israelites out from among them. (Exodus 7:4–5, KJV)

This is quite the moral and tactical puzzler. It's a bit like a chess grandmaster playing both sides of the board and saying, "I'm going to make black win, and just to show how great I am, I'll make some really dumb moves when I'm controlling white."

God's next move comes as quite a surprise. If it were you or me, and we had the Nile handy, we might go for a plague of crocodiles. That would shake things up. If it were Monty Python, on the other hand, the next bit might go something like this:

> And the Lord spake unto Moses, Go unto Pharaoh, and say unto him, Thus saith the Lord, Let my people go, that they may serve me. And if thou refuse to let them go, behold, I will smite all thy borders with frogs. And the river shall bring forth frogs abundantly, which shall go up and come into thine house, and into thy bedchamber, and upon thy bed . . . (Exodus 8:1–3, KJV)

Just to make sure the stunt doesn't fizzle, the Lord makes poor Pharaoh a frog-o-phobe. That soon becomes clear, 'cause croakers in his bed have him ready to throw in the linen:

> Then Pharaoh called Moses and Aaron, and said, "Pray to the Lord to

take away the frogs from me and my people, and I will let the people go to sacrifice to the Lord." (Exodus 8:8, KJV)

Show's over, right? Not quite. God zaps all the frogs, but no sooner are they piled up in the streets than Pharaoh scotches the deal. Back to work, Hebrews! Tote that barge, lift that 10-ton block!

So, continuing his glide down the scale of impressive feats, God next sends a plague of lice. Considering hygiene standards back then, we might wonder how anyone noticed the difference. Lice were not exactly rare. But things get sillier still: next, there's a plague of flies.

Seriously, flies? Anyone who's ever been within ten miles of a feedlot knows what a plague of flies is like. Irritating as hell, but nothing you can't handle. A swarm of them is hardly a miracle. Some scholars think that the text refers not to flies but to "wild animals" of some sort. The key thing we know about them is that they made up a swarm. Presumably, then, this was not a herd of elephants, nor a cackle of hyenas, a busyness of ferrets, or a convocation of eagles.

Grasshoppers swarm, as do ants, gnats, and butterflies. We can probably eliminate the last of these from our list of suspects. Who feels plagued by a swarm of Monarchs? The invaders can't be locusts, for reasons that will be apparent in a moment. For now, grasshoppers and ants are our best bets, as they are known to swarm and destroy agriculture and most everything else edible in their paths. However, once again we're flummoxed by the choice of a routine natural event. The existence of the verb "to swarm" in the original Hebrew, as well as in English, testifies to how regular it is. Unique events don't get their own verbs. Contrast the use of "swarm" with, say, the Big Bang, where we can only use distant verbal cousins like "explode" or "burst" to attempt to describe what we think happened when our universe came into being.

The only miraculous aspect of the story is that the Land of Goshen, home to the Israelites, remained unmolested. Now, for European Jews being forced to live in a ghetto had tragic consequences, culminating in the Holocaust. But back in merry olde Egypt, isolation sometimes had its advantages. Goshen is thought to have been located well to the east of the Nile, nearly halfway to the Red Sea. According to Genesis 45, it was granted to Jacob's family as they were fleeing famine. Goshen seems to have been a kind of reservation, although on better land than white America ever gave the American Indians.

The idea was clearly to keep the Jews from mixing with their "betters" in Egypt. No Egyptian would want to live in Goshen, and no Jews were allowed to squat in Egypt proper. That much is apparent in the final lines of the next chapter:

> "When Pharaoh calls you, and says, 'What is your occupation?' you shall say, 'Your servants have been keepers of livestock from our youth even until now, both we and our ancestors'—in order that you may settle in the land of Goshen, because all shepherds are abhorrent to the Egyptians." (Genesis 46:33–34, KJV)

Or, as the King James version puts it, an abomination. You know, like shellfish and gays. Maybe the shepherds had lice.

Actually, seems that "shepherds" is just a code word for "Jews," like "inner city" for "blacks" in contemporary American politics. If you riffle back a few pages to Genesis 43:32, you find the real target: "the Egyptians might not eat bread with the Hebrews; for that is an abomination unto the Egyptians." It's the table company not the matzoh, right?

At any rate, whatever swarm attacked Egypt, the Hebrews got off light. But still their bondage continues, for at least another plague or two.

6

 Microbial Plagues Aplenty

When, as foreordained, the lice and, let's assume, ants fail to move Pharaoh to free his Hebrew slaves, God ups the ante. The next plague kills off all the Egyptian livestock: horses, donkeys, goats, sheep, you name it:

> And on the next day . . . all of the livestock of the Egyptians died, but of the livestock of the Israelites not one died. Pharaoh inquired and found that not one of the livestock of the Israelites was dead. But the heart of Pharaoh was hardened, and he would not let the people go. (Exodus 9:6–7, NRSV)

Again, segregation has its advantages. When an infectious disease sweeps through a population, the best defense is quarantine. That can be very difficult to achieve, but the more virulent (fast-acting) a disease, the more rapidly an outbreak dies out, because the vectors (i.e., infected animals) die too quickly to spread the disease very far. This is a reason why, historically, Ebola has never traveled very far. The rise of major urban centers, jet transport, and the anti-vaccine rumor mill changes all that. Back then, however, all travel was on foot, so virulent plagues tended to be self-limiting.

What killed the livestock? There are numerous candidate diseases, but a veterinarian I consulted says that anthrax is a best bet. As anyone who remembers the dark days after 9/11 knows, it's a disease that can hide out in spore form to be transmitted long distances sealed in a letter, or carried about in a fleece. Spores are like the deep-freeze hibernation capsules for interstellar transport that you see in sci-fi movies—only much smaller and tougher.

Another possibility is one that Genghis Khan made inadvertent use of: the Rinderpest virus. It's a promiscuous pathogen that will infect not only sheep, goats, buffalo, and all other kinds of cattle but even your pet giraffe.

If so, the gap between Egypt and Goshen may have been just enough to keep the flock of the Hebrews pestilence free, while all the animals belonging to the Egyptians died. Well, maybe not all of their livestock. Curiously, just a hop, skip, and a verse later, the Lord sends a plague of boils upon the Egyptians and their animals! Don't take my word for it; harken to the Word of the Lord:

> Then the Lord said to Moses and Aaron, "Take handfuls of soot from the kiln, and let Moses throw it in the air in the sight of Pharaoh. It shall become fine dust all over the land of Egypt, and shall cause festering boils on humans and animals throughout the whole land of Egypt." So they took soot from the kiln, and stood before Pharaoh, and Moses threw it in the air, and it caused festering boils on humans and animals. (Exodus 10:8–10, NRSV)

Now, boils are an affliction we might today recognize as a microbial disease, such as chickenpox, caused by the Varicella virus, or a staph infection, courtesy of the *Staphylococcus aureus* bacterium. Staph is a common cause of boils, and is something both animals and humans contract—though the symptoms are different. For one thing, unless you shave them it's hard to spot boils on a cat or a goat. But in any event, it's certainly possible for an epidemic to hit both humans and livestock. The flu is a scary example, but fortunately the strains vary so much that most of the time only one species gets hit hard. Avian flu is a case in point: it kills millions of birds and typically a few human handlers, but no version has been known to go airborne and infect both birds and humans simultaneously. Knock wood.

Of course, if we assume that God has his own gene-splicing setup, then all bets are off. He could undoubtedly design a chimera virus or a suite of viruses that would kill any combination of animals. That's well within the capability of bioweapons labs today, along with the ability to design viruses to remain inactive until a certain trigger—say, Egyptian cotton?—detonates the plague. Let's just hope someone remembers to insert an off-switch into the code.

7

 Grinding It Out

From boils it's on to crop-destroying thunderstorms and hail. Frankly, these are so nonmiraculous they don't deserve discussion. Pharaoh wasn't impressed, and no farmer would be either. It's what happens when you fail to buy crop insurance.

Next comes another swarm: locusts. Now, the one really interesting thing about locusts, other than their ravenous appetites, is timing. Various species have evolved a clever strategy: they go dormant for years—ten, thirteen, even as much as seventeen years at a stretch—and then burst out of the ground in huge swarms. No one knows just why. It may have begun as a way to survive cyclical droughts, or it may be a way to ensure that this particular species overwhelms all others in its chosen year of swarming. One thing's for sure: it's murder on agricultural societies. The Bible offers a vivid and horrifying description:

> They covered the surface of the whole land, so that the land was black;
> and they ate all the plants in the land and all the fruit of the trees that
> the hail had left; nothing green was left, no tree, no plant in the field,
> in all the land of Egypt. (Exodus 10:15, NRSV)

That's a plague that many a farmer down the ages would recognize. It continues today. In March 2013 a plague of locusts made its way from Egypt to Israel . . . just as Passover began. Payback? No one seems to have taken it that way. Will miracles never cease?

After the locusts have moved on, a time of darkness sets in. Now, I know what you're thinking: doesn't that happen every day? Sure. But this is an Unusual Darkness—one that lasts three days. Solar eclipses only last minutes, so we know it can't have been one of those. This particular darkness has a special quality that helps point us toward an explanation. It is darkness you can touch. I told you it was particular.

> Then the Lord said to Moses, "Stretch out your hand toward heaven so that there may be darkness over the land of Egypt, a darkness that can be felt." So Moses stretched out his hand toward heaven, and there was dense darkness in all the land of Egypt for three days. (Exodus 10:21–22, NRSV)

One possibility, obviously, is a thick fog. But perhaps that's not dark enough, and besides fog is rare in Egypt's hot climate. Smoke from wildfires could do the job, but unlike the Western United States, the Sahara lacks scrub vegetation, let alone forests, to burn. Another possibility is volcanic haze. As noted in the Sodom and Gomorrah chapter, there are live volcanoes in the Middle East. A powerful eruption can put millions of tons of fine debris into the atmosphere, darkening the skies and falling to earth as a rain of dust you can reach out and touch.

But let's buy local. There's a phenomenon right at hand in Egypt that can easily explain a three-day palpable darkness: the sandstorm. As a boy living in Cairo (a little more recently than Pharaoh), I experienced sandstorms myself. The late spring winds, blowing across the Sahara, bounce grain against grain, creating a negative static electric field that causes the grains to repel each other and aloft they go. Eventually, a vast wall of darkness moves across the face of the earth, shrouding everything in its path. Once engulfed in a sandstorm, you can feel, taste, and see the darkness. Moses probably didn't stand out there savoring it for long. A sandstorm stings!

Of course, the Pharaoh, being a local boy himself, is not so much impressed as he is fed up with this tiresome show. His streets are full of stinking animal carcasses, he's got welts all over, and what little grain he has left is full of grit. In short, it was life as usual in Ancient Egypt.

Things were tough in those days. Not even pharaohs, whom the Egyptians treated as living gods, had an easy time of it. A study of some

3,000 mummies published in the *Journal of Comparative Human Biology* shows that upper-class Egyptians suffered from a torture chamber's worth of diseases: parasites, middle-ear infections, bone-on-bone osteoarthritis, and most common of all, even among the living gods, pulsating abscesses in the jaw, caused by tooth enamel lost to that ever-present grit in their food.

So it would have been tough to impress old Pharaoh with hardship—even if God hadn't fixed the match by "hardening Pharaoh's heart," as the Bible repeatedly tells us. At long last, nine plagues into this dreary contest of wills, Pharaoh is ready to call it quits. But this being the Middle East, he can't resist the urge to haggle over the End Game.

> And Pharaoh called unto Moses, and said, Go ye, serve the Lord; only let your flocks and your herds be stayed: let your little ones also go with you. And Moses said, Thou must give us also sacrifices and burnt offerings, that we may sacrifice unto the Lord our God. Our cattle also shall go with us; there shall not an hoof be left behind; for thereof must we take to serve the Lord our God; and we know not with what we must serve the Lord, until we come thither. (Exodus 10:24–26, KJV)

Now, this is where Pharaoh is supposed to jump in and say, "Moses, my friend, I like the cut of your beard, so I tell ya what I'm gonna do. I'm gonna let ya take half your flock! Deal?" I mean, that kind of thing has been going on in the Middle East since sand was invented. But God has other ideas. With a grand finale in mind, he puts the kibosh on any pact.

> But the Lord hardened Pharaoh's heart, and he would not let them go. And Pharaoh said unto him, Get thee from me, take heed to thyself, see my face no more; for in that day thou seest my face thou shalt die. And Moses said, Thou hast spoken well, I will see thy face again no more. (Exodus 10:27–29, KJV)

So, it's on to the climactic, show-stopping plague. Apparently, though, God needs to raise a little capital for the finale. It's not clear why, but perhaps he needed money to buy his priests some new robes. Judge for yourself:

And the Lord said unto Moses, Yet will I bring one plague more upon Pharaoh, and upon Egypt; afterwards he will let you go hence: when he shall let you go, he shall surely thrust you out hence altogether. Speak now in the ears of the people, and let every man borrow of his neighbour, and every woman of her neighbour, jewels of silver and jewels of gold. And the Lord gave the people favour in the sight of the Egyptians. Moreover the man Moses was very great in the land of Egypt, in the sight of Pharaoh's servants, and in the sight of the people. (Exodus 11:1–3, KJV)

God then fills Moses in on the details of the plan—it's a plan that amounts to nothing less than genocide. Not indiscriminate genocide, however. It targets only firstborn boy babies. Nice, huh? In order not to have the Hebrews suffer the same fate, God says they must kill year-old male lambs and smear the blood on their doorframes. Good symbolism . . . if you're planning a slaughter of the innocents.

8

 Great Time to Be a Younger Sibling

Having stage-managed the preludatory plagues, it's time for God to appear on stage himself, as he explains to Moses:

> I will pass through the land of Egypt this night, and will smite all the firstborn in the land of Egypt, both man and beast; and against all the gods of Egypt I will execute judgment: I am the Lord. And the blood shall be to you for a token upon the houses where ye are: and when I see the blood, I will pass over you, and the plague shall not be upon you to destroy you, when I smite the land of Egypt.
> (Exodus 12:12–13, KJV)

And, shazam! Death and sorrow strike every last household in Egypt. Pharaoh, as foreordained, summons Moses and tells him to take his people and his flocks and skedaddle.

Can such a thing be explained? Well, singling out firstborn offspring is something nature does not normally do. People, however, are another matter. In many societies, the problem of diluting property holdings down the generations has been solved by privileging the firstborn male. The ancient Israelites were certainly one such society, and this gives rise to a possible explanation. Perhaps a deadly fungus infected such grain as was saved from the torrential storms that flattened it—a not improbable turn of events for damp grain. In that case, the firstborn sons, being the privileged ones, would have had the biggest helping. Being young, they also would have been particularly vulnerable to the toxins.

70

This only works if we're not too literal. Childhood mortality was an everyday experience for the ancients—indeed for all of humanity except the last few generations, and only those lucky enough to live in advanced societies. The story of the Tenth Plague is noteworthy only because it picked out an unusual pattern from among routine childhood deaths. Until quite recently, women had children in rapid succession. So, it's plausible that a firstborn would be young enough to be susceptible to tainted grain yet also have a younger sibling. Suckling babes would either have been spared if their mothers' bodies suppressed or filtered out whatever afflicted the weaned children. Still, it's unlikely that such a calamity would have killed firstborns exclusively. At most, it would have created a noticeable trend that turned into legend.

That same caution pertains to all the plagues. Taken literally, the stories cannot be explained by science, because each plague explicitly includes planning and execution by God. However, if we view them as whisper-down-the-lane accounts of actual experiences that were later appropriated by the authors of the Bible, then there is nothing inexplicable here. Well, almost nothing.

We are left with the mystery of how God could do such a thing— or, more precisely, how anyone could believe that God could do such a thing. The Bible, we are told, is God's Word, and God, we are told, is all knowing, all powerful, and all good. But to deliberately bring about a genocide is awful; to do it for the sole purpose of showing off is monstrous. Yet, shortly before killing all the firstborns, that's what God gives as his purpose: "The Lord said to Moses, 'Pharaoh will not listen to you, in order that my wonders may be multiplied in the land of Egypt.'" (Exodus 11:9, NRSV)

The wonder is that anyone could have believed such a God was good. No wonder many modern Christians believe that Exodus speaks to us not of literal events—as we'll shortly see, the mass escape from Egypt appears to be a myth—nor of eternal truths. Rather, it is a faded, crinkled sepia snapshot of a time when natural disasters could only be understood as magical, and moral decency could only stretch to kinfolk.

But, as horrific as this tableau may be, the drama's not yet done. Everybody's favorite scene from the 1956 Hollywood epic *The Ten Commandments* has yet to be enacted.

9

 Parting of the Red, or Reed, Sea

After the supernatural slaughter of all the Egyptian firstborns in Egypt, Pharaoh finally relents. He might have done this much earlier, but God has been "hardening his heart," apparently to keep the spectacle going. At long last, however, Pharaoh says, go on, take your beasts with you, and worship your god. Then he adds, "and bless me also"—or at least that's how the Bible writers interpreted his words. If it were me, after all those frogs and lice and flies, oh, and the dead babies and whatnot, I would probably have said something unprintable. Anyway, with all the Hebrew cheering it was hard to hear exactly what Pharaoh said. You decide.

With God's connivance, the Israelites "borrow" silver, gold, and fine clothing from the Egyptians and then skip town. I'm not making this up. Here's the Bible again: "the Lord had given the people favor in the sight of the Egyptians, so that they let them have what they asked. And so they plundered the Egyptians." (Exodus 12:36, NRSV)

This pilfering is but a prelude to the land grabs ahead. In exchange for observing Passover, the Israelites are bound for "the land of the Canaanites, the Hittites, the Amorites, the Hivites, and the Jebusites . . . a land flowing with milk and honey." In short, the Promised Land awaits.

But not quite yet:

When Pharaoh let the people go, God did not lead them by way of the land of the Philistines, although that was nearer; for God thought, "If the people face war, they may change their minds and return to Egypt." So God led the people by the roundabout way of the

wilderness toward the Red Sea. . . . The Lord went in front of them in
a pillar of cloud by day, to lead them along the way, and in a pillar of
fire by night, to give them light, so that they might travel by day and by
night. (Exodus 13:17–22, NRSV)

The "pillar of cloud by day and pillar of fire by night" are surely
suggestive. Together, they are a good fit with an erupting volcano. If an
eruption occurred in the Red Sea, it would produce exactly the kind of
column of cloud by day and pillar of fire by night the Bible describes.
But what are the odds?

As it happens, there's a tectonic rift right under the Red Sea, and
since 2007 it has been putting on a show of biblical proportions. A series
of volcanic eruptions culminated in the birth of a new island, towering
100 meters above sea level. Wenbin Xu and Joël Ruch, researchers in
Sigurjón Jónsson's lab at the King Abdullah University of Science and
Technology, have had a front-row view.

The Arabian Peninsula is gradually swinging away from Africa,
but it doesn't always proceed smoothly. The Saudi-based researchers
detected a series of earthquakes shortly before the eruptions, leading
them to conclude that the southern Red Sea is undergoing a "rifting
episode," in which diverging tectonic plates jerk past each other. "It's like
you keep pulling on an elastic band," Jónsson told the journal *Nature Asia*.
"At some point, it snaps, and that's when you have a rifting episode."

Back in the time of Moses, the Lord, who launched this Red Sea
tour, decides to stir up some mischief. Once again, God is going to mess
with Pharaoh's mind to make him do something really, really stupid.

Then the Lord said to Moses: Tell the Israelites to turn back and camp
in front of Pi-hahiroth, between Migdol and the sea, in front of Baal-
zephon; you shall camp opposite it, by the sea. Pharaoh will say of the
Israelites, "They are wandering aimlessly in the land; the wilderness
has closed in on them." I will harden Pharaoh's heart, and he will
pursue them, so that I will gain glory for myself over Pharaoh and all
his army; and the Egyptians shall know that I am the Lord. And they
did so. (Exodus 14:1–4, NRSV)

God has turned the desert into a vast colosseum where a murderous show will soon take place. The Hebrews will get a big scare, but for them it's all in fun. The Egyptians haven't got a chance. By today's humanitarian standards, this toying with the Pharaoh and his minions seems downright cruel.

But, lest we judge too harshly, consider what passed for entertainment in ancient Rome. Here's an account by the Roman philosopher and writer Lucias Anneus Seneca:

> The combatants have no protective covering; their entire bodies are exposed to the blows. No blow falls in vain. This is what lots of people prefer to the regular contests, and even to those which are put on by popular request. And it is obvious why. There is no helmet, no shield to repel the blade. Why have armour? Why bother with skill? All that just delays death. In the morning, men are thrown to lions and bears. At mid-day they are thrown to the spectators themselves. No sooner has a man killed, than they shout for him to kill another, or to be killed. The final victor is kept for some other slaughter. In the end, every fighter dies.

The Pharaoh and his forces have no more choice than the prisoners of Rome, and their doom is no less certain. But, the show must go on:

> Moses said to the people, "Do not be afraid, stand firm, and see the deliverance that the Lord will accomplish for you today; for the Egyptians whom you see today you shall never see again. The Lord will fight for you, and you have only to keep still." Then the Lord said to Moses, "Why do you cry out to me? Tell the Israelites to go forward. But you lift up your staff, and stretch out your hand over the sea and divide it, that the Israelites may go into the sea on dry ground. Then I will harden the hearts of the Egyptians so that they will go in after them; and so I will gain glory for myself over Pharaoh and all his army, his chariots, and his chariot drivers. And the Egyptians shall know that I am the Lord, when I have gained glory for myself over Pharaoh, his chariots, and his chariot drivers." (Exodus 14:15–18, NRSV)

The stage directions become a little confusing here, with the pillar of cloud shifting about, but a familiar story emerges:

> Then Moses stretched out his hand over the sea. The Lord drove the sea back by a strong east wind all night, and turned the sea into dry land; and the waters were divided. The Israelites went into the sea on dry ground, the waters forming a wall for them on their right and on their left. The Egyptians pursued, and went into the sea after them, all of Pharaoh's horses, chariots, and chariot drivers. At the morning watch the Lord in the pillar of fire and cloud looked down upon the Egyptian army, and threw the Egyptian army into panic. He clogged their chariot wheels so that they turned with difficulty. The Egyptians said, "Let us flee from the Israelites, for the Lord is fighting for them against Egypt.". . .
>
> Then the Lord said to Moses, "Stretch out your hand over the sea, so that the water may come back upon the Egyptians, upon their chariots and chariot drivers." So Moses stretched out his hand over the sea, and at dawn the sea returned to its normal depth. As the Egyptians fled before it, the Lord tossed the Egyptians into the sea. The waters returned and covered the chariots and the chariot drivers, the entire army of Pharaoh that had followed them into the sea; not one of them remained. (Exodus 14:21–29, NRVS)

Is this explicable in natural terms? Well, not entirely, to be sure. If God actually peeps out from a pillar of cloud and fire, then there's no explaining it. But if we leave out the flourishes and take the essence of the story, we have at least two shots at making sense of it.

Even without the supernatural frills, however, the challenge is formidable. The Red Sea, with the help of the modern Suez Canal, cleaves Africa from Asia. It should be no surprise, then, that it is both steep and deep. On average, it is more than 1,600-feet deep, plunging to a nadir of nearly a mile and a half. At its widest point, it gapes more than 220 miles. This presents two problems: first, how to move a huge volume of water to make safe passage for the Israelites, and second, how to get them across a V-shaped chasm.

The surprising answers come from a surprising source: a Canadian apologist. Writing at the Interactive Bible Web site, apologist Steve

Rudd offers a startling suggestion: the crossing took place not at what we typically think of as the Red Sea, but on the other side of the Sinai Peninsula, near the tip at what today is the resort area of Sharm el Sheikh. (By a nice coincidence, "Rudd" in its Old English means "red," as in Red Sea.) The actual point of crossing, he claims, is a strait at the mouth of the Gulf of Aqaba. His argument is too long and choppy to quote in full, but here's a taste of his passionate, enumerated style, followed by his favored solution. Rudd first disputes claims that either the Israelites crossed at the northernmost tip of the Suez Gulf, where the canal now connects with the sea, or they crossed one of the chain of lakes that used to dot the land between the gulf and the Mediterranean prior to the construction of the canal.

1. When Israel saw the Egyptian army getting ready to attack them: "Then they said to Moses, "Is it because there were no graves in Egypt that you have taken us away to die in the wilderness? Why have you dealt with us in this way, bringing us out of Egypt? Is this not the word that we spoke to you in Egypt, saying, 'Leave us alone that we may serve the Egyptians'? For it would have been better for us to serve the Egyptians than to die in the wilderness."" (Exodus 14:11–12)

2. This is where any suggested crossing point like the Bitter Lakes or the northern Suez becomes plain silly because they are just too close to Egypt to say this. They are clearly NOT IN THE WILDERNESS, since the Bitter Lakes are about 25 miles from the edge of Goshen where they lived. The Northern Suez crossing is only 60 miles. Far too close to worry about dying in the wilderness if it was just a day's walk back to your old bed in Goshen. Why they would probably walk to the Bitter Lakes to fish on their day off just for fun.

Having made the case for a 250-mile trek down the Sinai Peninsula, Rudd then stakes his claim for a crossing into what is today Saudi Arabia:

1. The Gulf of Aqaba is a very deep channel of water ranging from 800–1800 meters in the middle. However at the Straits of Tiran, there is a natural land bridge so the deepest point the Israelites would encounter is only 205 meters.

2. The crossing at the Straits of Tiran is 18 km long and a natural land bridge provides for an 800 meters wide pathway the full distance of the crossing. The Straits of Titan have a shallow coral reef in the middle with a one-way shipping lane on either side. From modern nautical charts, we can see that the eastern "Enterprise Passage" is 205 meters deep and 800 meters wide and the western "Grafton Passage" is only 70 meters deep and 800 meters wide. A diver need go only 13 meters at deepest point on top of Jackson's Reef from the surface.

3. Coral growth over the last 3500 years since the miraculous crossing means that we cannot really know what the sea floor looked like exactly back then. For example, as the coral grew up and came to the surface, the tides flowing around the coral would dig a natural channel deeper on the north and south ends of the reef where all the water would flow around. Gradually, the coral reefs would act like a partial dam over the center 80% of the strait. This is a very realistic scenario and means that 3500 years ago, the coral was under water and therefore the tides would not dig the deep channel at either end of the reef where it is today.

4. But even with the depths we see today, it causes no problems for the exodus crossing. The slope of descent is far more important than the depth. The Straits of Tiran, as we see them today pose absolutely no problem for a crossing by a million people since the slope is shallow and the depth is no more than 600 feet (205 meters).

There's a charming confidence in the writings of apologists. And no wonder: since the outcome (confirmation of what they understand the Bible to say) is foreordained, all they need do is offer plausible-sounding descriptions of what can be so described, and leave the rest in God's hands.

So there you have it: a "very realistic" scenario in which the Israelites cross the parted Red Sea with "absolutely no problem" via a coral-encrusted land bridge. No word on how a million refugees, weighted down with stolen silver and gold, managed to water themselves and their flocks on what must have been at least a ten-day trek through arid wilderness.

That population number, by the way, comes from textual interpretation. Astute readers will recall that the Pharaoh's ire was first aroused by the excessive Hebrew breeding. In another page on the same Web site, Rudd devotes himself to estimating the number of Israelites who crossed the Red Sea. Guess what: a million is his low-end estimate. The Bible only counts men, he notes. Given that starting point, he figures there were 600,000 men on the move. If every man had a woman, and every woman had some kids . . . let's see . . . carry the three and . . . well, here's his conclusion: "We do not believe that an exodus population of 2.5 million defies human reasoning in any way. It is a very believable number."

Okay, we've upped the ante. Odd, then, that the Bible doesn't speak of the truly miraculous furnishing of at least 2.5 million gallons of water a day for the refugees, plus water for the animals, in the midst of a desert. To bring that much water along on a 10-day trek, the Israelites would have had to drag approximately 50 Olympic swimming pools filled with fresh water behind them.

But leaving that aside, let's consider what could have exposed the submerged land bridge at the spot Rudd identifies. Among other explanations, the Bible says a persistent wind parted the waters. That won't do for a passage at least 70 meters deep. If you imagine a football field turned into a swimming pool that's more than two-thirds as deep as the field is long, you can imagine that not even a hurricane would blow all the water out of it. Liquid water is massive. Later in this chapter, we'll see that at most wind phenomena can shift water only several meters in depth. To switch measures, a hurricane storm surge will at most displace a 20-foot column of water, nowhere near the 230-foot depth of Rudd's land bridge.

There is, however, another natural force that could play a role, and it happens to align with the volcanic eruption we posited earlier. When tectonic plates suddenly shift past each other, we get an earthquake. Volcanic eruptions are often accompanied by such shifts. If it happens that the plates move vertically as well as laterally, and if they happen to lie under an ocean, an enormous amount of water will move as well.

When it does, a tsunami may follow. "Tsunami" is a Japanese word meaning "giant wave." It is appropriate, therefore, that we take a look at the worst natural disaster Japan suffered in the twentieth century: the great Kanto earthquake of 1923.

The sudden upsurge of the continental plate displaced millions of gallons of seawater, which formed into a great wave that headed straight for the bay's cove-sculpted shoreline. The tsunami began as an almost imperceptible swell, no more than three feet high: fishing boats near Oshima island at the time the earthquake struck reported a minor disturbance, nothing more. Though unseen and barely felt by those at sea, the wave raced across Sagami Bay at a speed approaching 600 miles an hour.

As the swell approached the beaches it brushed against the shallowing sea floor, which slowed its movement and drastically changed its shape. Like a piece of carpet shoved suddenly against the wall, the wave curled upward, growing to terrifying size. The precise height of the wave when it hit the shore was determined both by the depth of the sea floor and the shape of the coastline. The narrower the inlet the greater the tsunami size. At Atami the wave squeezed through a slender cove and reached the height of 35 feet, swamping the town minutes after the shock and drowning 300 people.

The scene was repeated in many other places, with appalling consequences. More than 100,000 people died—some put the total at nearly 150,000—and the entire Kanto plain, encompassing Tokyo, Yokohama, and many other cities were in ruins.

Could this have happened in the Holy Land? To his credit, Rudd admits that no one is sure of where any of these places were. There were no maps and the archaeological evidence . . . well, we'll come to that presently.

For the moment, the biggest problem with Rudd's scenario is that scholars are pretty much unanimous in concluding that the Bible does not refer to the Red Sea. The translation is faulty. The original text, scholars tell us, refers to the "Reed Sea." It is remarkable that in both Hebrew and English the words are quite similar. Still, nothing miraculous in that.

The "reed sea" is likely one of the marshy lakes that used to dot the line now occupied by the Suez Canal. Gary Byers, writing in the journal *Bible and Spade*, notes:

There is general agreement among scholars today, both liberal and conservative, that yam suph means "Reed Sea." The Hebrew suph definitely referred to a water plant of some sort, as indicated in Exodus 2:3–5 and Isaiah 19:6–7, where reeds in the Nile River are mentioned. In fact, it is probable that the Hebrew suph ("reed") is an Egyptian loanword—from the hieroglyph for water plants.

Two reedy marshes remain: the Great Bitter Lake and Lake Timsah. How could a lake fit into the story? Well, it depends on what version you adopt. Scholars say that at least three, and possibly four, independent narratives have been interleaved in the Book of Exodus. Just as the Creation story in Genesis comes in two strikingly different accounts, here we see several different causes parting the Red Sea.

In one, Moses parts the Red Sea (with God's help) by stretching out his hands over the waters, abracadabra-style. That's the version we see in the 1956 Cecil B. DeMille Hollywood blockbuster starring Charlton Heston as Moses.

But the text of Exodus offers a different tale: "The Lord drove the sea back by a strong east wind all night, and turned the sea into dry land; and the waters were divided." This lends enough scientific plausibility to the account to encourage some to think it through in natural terms. No one has done so more thoroughly than engineer Carl Drews. A Christian who works for the National Center for Atmospheric Research, Drews has devoted a master's thesis to explaining how a weather event could have led to the parting of the sea.

As recounted by science writer Chris Mooney in the *Washington Post*, Drews' scientific scenario passed peer review and, while controversial, has come to be regarded as a plausible account of how the waters might have parted.

Drews and coauthor Weiqing Han speculate that around 1250 BCE Lake Tanis ballooned out from a branch of the Nile in the delta, just shy of the Mediterranean. They hypothesize that at the neck of the lake, where it joins the Nile, the crossing took place. How?

The researchers ran a series of computer simulations of a rare but well-established weather event called a "wind setdown." When a strong storm is offshore, it sucks water up in a storm surge—something coastal residents of the southern United States know all too well. But a storm can also blow water away on its periphery. When that happens, the "push" can sometimes expose an underwater ridge.

However, to explain the Bible version of the Great Escape, the winds can't be too strong and the ridge can't be too deep. Anything more than about 60 mph, and the winds would make a crossing by thousands of people impossible. Anything deeper than a couple of meters, and the wind would be unable to part the waters.

Working from that knowledge, Drews and Han drew up a scenario that offers a reasonable fit with the "reed sea" crossing. They studied what the topography of the eastern delta of the Nile would have been like thousands of years ago and settled on a reconstruction of the geography. The Lake of Tanis, Drews tells Mooney, "was a shallow brackish lagoon, and that was the ideal place for these papyrus reeds to grow. So if you want to find a sea of reeds . . . that's it."

Having settled on a location far, far away from the one Steve Rudd claims to have identified, Drews and Han went to work on a Bluefire supercomputer at the National Center for Atmospheric Research. In their scenario, 60 mph winds blow continuously from the east for twelve hours along a narrow portion of the lagoon called the Kedua Gap, and then, *seem, seem, alameem*!

> A dry, traversable gap in the waters opens at 9:36 hours. It appears feasible for a large number of people to make their way across the exposed mud flats. Maybe not millions, but a hefty crowd. The midpoint of the land bridge is at (30.9812° N, 32.4553° E). The passage is about 5 km wide initially, and it later expands up to 6 km wide. This land bridge remains continuously open until 13:30 hours, leaving 3.9 hours for the company to cross the Kedua Gap.

It's an ingenious and well-researched solution. The lagoon fits well with the biblical account of the Pharaoh's chariots becoming stranded in the mud. No one has poked holes in the science. Indeed, such wind setdowns have been observed to clear a passage through water before, and there's even been something like it recorded in nineteenth-century Egypt.

There are only two hitches. One, the reconstruction of the geography is highly speculative—so much so that, given the supposition of a wind setdown, finding a solution was inevitable. If you have a key and you get to carve the keyhole, the key will always fit.

The other, an even bigger obstacle to a scientific case, is this: no good evidence exists to support the Exodus story. It now appears likely that the Israelites were not enslaved en masse in Egypt. Sorry, Cecil. In all probability, the flight from Egypt never happened, and Pharaoh never gave chase.

Archaeologist Stephen Gabriel Rosenberg, writing in the *Jerusalem Post*, bemoans the burden of his profession as mythbusters:

> The whole subject of the Exodus is embarrassing to archaeologists. The Exodus is so fundamental to us and our Jewish sources that it is embarrassing that there is no evidence outside of the Bible to support it. So we prefer not to talk about it, and hate to be asked about it.
>
> For the account in the Torah is the basis of our people's creation. . . . So that makes archaeologists reluctant to have to tell our brethren and ourselves that there is nothing in Egyptian records to support it. Nothing on the slavery of the Israelites, nothing on the plagues that persuaded Pharaoh to let them go, nothing on the miraculous crossing of the Red Sea, nothing.

As a consolation prize, Rosenberg dreams up a scenario in which Pharaoh Akhenaten might have deployed skilled Hebrew slaves to make mud bricks for the construction of a new temple city, but he admits that he has no evidence to back it.

Rather than seeking keyholes to fit keys—that is, only looking for evidence to validate Bible stories, as past generations of Holy Land investigators did—contemporary archaeologists allow the evidence to furnish the tale, and a stunningly different story it tells.

Ze'ev Hertzog, a professor at Tel Aviv University, was among the first to break with the apologetic tradition of archaeology. He was roundly attacked by traditionalists for boldly conceding an unwelcome truth: "Following 70 years of intensive excavations in the Land of Israel, archaeologists have found out: The patriarchs' acts are legendary, the

Israelites did not sojourn in Egypt or make an exodus, they did not conquer the land."

Subsequently, another Israeli archaeologist, Ammon Ben-Tor, made a discovery that has given rise to a revolutionary idea. Ben-Tor led an excavation that found clear evidence of fiery destruction in the Canaanite city of Hazor. The Egyptian records show no expeditions to Canaan, so Ben-Tor concludes that only the Israelites could have done it. Perhaps, he thought, a civil war erupted. His protégé, Sharon Zuckerman, took the idea further. Noting that the pattern of destruction does not include signs of a military battle, she concludes that an uprising rather than an invasion took place.

In sum, the evidence shows no sign of Hebrew enslavement, no escape from Egypt, no wandering in the wilderness, and no conquest in Canaan. Instead what it points to—albeit inconclusively—is a revolution, followed by the rise of a myth of justification. You know . . . like, "God gave this land to me."

10

 Manna from Heaven

After their supposed escape, the Israelites are bound for the Promised Land. But there's a cry in the wilderness, and it goes something like this: "Are we there yet?"

It's a fair question. They've been on the road for 45 days now. Not that there's even a road. The Israelites, the Bible tells us, are stumbling about in "the wilderness of Sin, which is between Elim and Sinai." Not so much as a Flying J Travel Plaza out there. The Chosen People are thoroughly sick of it. They have one big *kvetch*: food.

> The whole congregation of the Israelites complained against Moses and Aaron in the wilderness. The Israelites said to them, "If only we had died by the hand of the Lord in the land of Egypt, when we sat by the fleshpots and ate our fill of bread; for you have brought us out into this wilderness to kill this whole assembly with hunger." (Exodus 16:1–3, NRSV)

By now Moses is probably wishing he had kept his day job as a prince in Pharaoh's palace. Not to worry, though, God's got a plan: he's going to air-drop relief supplies.

> Then Moses said to Aaron, "Say to the whole congregation of the Israelites, 'Draw near to the Lord, for he has heard your complaining.'" And as Aaron spoke to the whole congregation of the Israelites, they looked toward the wilderness, and the glory of the Lord appeared in the cloud. The Lord spoke to Moses and said, "I have heard the

complaining of the Israelites; say to them, 'At twilight you shall eat meat, and in the morning you shall have your fill of bread; then you shall know that I am the Lord your God.'" (Exodus 16:9–12, NRSV)

Now, this is truly interesting. Remember in the last chapter when Steve Rudd calculated the number of Israelites making this journey? Based on figures in the Bible, he calculated a minimum of a million, but a more likely figure of 2.5 million. So, "draw near" takes on spectacular meaning in this context. Imagine yourself at a game between the Ohio State Buckeyes and the Michigan Wolverines. There's a sellout crowd in Ohio Stadium—just over 100,000 people. At halftime, you are invited to come onto the field and address the crowd . . . only the public address system fails. You have to speak to 100,000 people in your natural-born voice. Can you make yourself heard? All the way to the upper decks?

If you scream your lungs out, the guys with the body paint and the beer in the nosebleed section will hear the sound—if everyone else keeps still. A typical stadium public address system needs to achieve 95 decibels for everyone to hear. The loudest scream ever recorded was by a woman who hit a shattering 129 decibels. (Decibels go up logarithmically, which means every ten-point increase is ten *times* louder.) But screaming is one thing, producing intelligible speech quite another. A whole stadium's roaring peaks at about 130 decibels, but you can't understand a damn thing they're saying.

Suppose everyone kept quiet and Aaron used his best outdoor voice. Even if Moses' big bro were gifted with the biggest voice and the best vocal articulation ever, in the Ohio Stadium he'd only be addressing one-tenth of the minimum estimated population of Israelites on the march.

Strain your imagination and grow that stadium by ten times. At 1,666 feet, it stands taller than the Empire State Building in an ever-expanding oval. What does it take to speak to a million people in such a gigantic stadium? You might think you need to boost your PA system by ten, but you'd be wrong. The crowd is effectively a two-dimensional layer blanketing the stadium, but sound swells in three dimensions, and that means it diminishes by the good old inverse square law.

So, when you increase the stadium's capacity by ten, you need to amp up the volume by 100. No wonder Moses said to his big brother,

"You go talk to them." Somehow, Aaron needs to exceed the sound of a jetliner at takeoff, the report of a howitzer firing shells, a thunderclap, and the roar of a Saturn V rocket, all rolled into one. Since sound waves turn into shock waves at just 194 decibels, when Aaron speaks, the front ranks of the Israelites are gonna shake, rattle, and roll . . . and die a horrible death.

But maybe there's an out. The Occupy Wall Street movement, that futile exercise in leaderless democracy, developed what it called the "human microphone" or "people's microphone." (No one had the authority to decide which term was correct.) When someone had something to say to the whole crowd, they would yell "mic check" and this would be repeated through the whole crowd. Thereafter, the speaker would utter a short phrase, which the crowd nearest would repeat in loud unison, and then the further ranks would repeat that until the phrase carried all the way to the end of the crowd. Of course, as anyone who ever played "Whisper Down the Lane" (or "Telephone") in school knows, there's a considerable risk of garbling the message along the way.

The British comedy troupe Monty Python picked up on that in their satirical movie *Life of Brian*. Happening upon the Sermon on the Mount, Brian and his mum stop to listen. With just several hundred people in front of them, they find it impossible to make out what Jesus is saying.

"Speak up!" yells Brian's crabby mother, but to no avail. Those around them try to catch and pass on the faint murmurs that reach them. "'Blessed are the cheesemakers,'" says one. Yea, verily . . . and blessed are the chic.

At least Jesus had the good sense to speak in short, catchy declarations. Despite facing a crowd of at least a million people, Moses elbows Aaron aside and addresses his people like a grouchy deli owner:

> [W]hat are we, that you complain against us? . . . When the Lord gives you meat to eat in the evening and your fill of bread in the morning, because the Lord has heard the complaining that you utter against him—what are we? Your complaining is not against us but against the Lord. (Exodus 16:7–8, NRSV)

Somehow or other, word gets about: the Lord is offering a meal plan. It includes a mysterious breakfast entree called "manna." Like the stuff itself, the word is enigmatic: it seems to mean "what is it?" The Bible offers several descriptions. As often happens in new restaurants, the initial wording on the menu is a bit careless, making manna sound pretty yucky:

> In the evening quails came up and covered the camp; and in the morning there was a layer of dew around the camp. When the layer of dew lifted, there on the surface of the wilderness was a fine flaky substance as fine as frost on the ground. When the Israelites saw it, they said to one another, "What is it?" For they did not know what it was. (Exodus 16:13–15, NRSV)

I seem to remember a similar experience at summer day camp. Maybe we got served manna. Ugh. Could it be? The consensus among us campers centered on marinated monkey meat. But what did we know? Anyone who's spent time in a train station infested with pigeons has also seen a fine, white flaky substance on the ground. Fine, yet gross.

But, come on. Not even Yahweh, with all his plagues, is going to give his Chosen People poop for breakfast. God has higher things in mind, like turning a meal into an obedience drill:

> This is what the Lord has commanded: "Gather as much of it as each of you needs, an omer to a person according to the number of persons, all providing for those in their own tents." The Israelites did so . . . they gathered as much as each of them needed. And Moses said to them, "Let no one leave any of it over until morning." But they did not listen to Moses; some left part of it until morning, and it bred worms and became foul. And Moses was angry with them. (Exodus 16:16–20, NRSV)

Okay, so maybe manna spoils fast. It's not unnatural that food deposited on the ground would provide a good place for flies and other insects to deposit their eggs. "Worms" can also be translated as "maggots," and in any case there were no biologists in those days to explain the difference. However, there is one miraculous-seeming feature of the spoilage: it takes a day off each week.

> On the sixth day they gathered twice as much food, two omers apiece. When all the leaders of the congregation came and told Moses, he said to them, "This is what the Lord has commanded: 'Tomorrow is a day of solemn rest, a holy sabbath to the Lord; bake what you want to bake and boil what you want to boil, and all that is left over put aside to be kept until morning.'" So they put it aside until morning, as Moses commanded them; and it did not become foul, and there were no worms in it. (Exodus 16:22–24, NRSV)

Aha. So, if you bake or boil the stuff, it doesn't spoil right away. Pretty much like anything else that you bake or boil, thereby killing any bacteria or eggs hidden away in it. But still . . . what is it?

Apart from being fine and flaky, the Bible tells us that "it was like coriander seed, white, and the taste of it was like wafers made with honey." Oh, and one more detail: "When the sun grew hot, it melted."

Scholars who have studied this have some ideas of what manna could be. One possibility: the resin of Tamarisk trees. Now, I know what you're thinking: trees? In a desert? You've got a point there, but scholars tell us that there used to be more of them, and remember this huge crowd must be finding water holes somewhere in the wilderness. If anywhere, that is where trees would be. So, resin is a possibility, but not an entirely satisfactory one. It's not white, and being essentially like gum, it's nothing you could describe as flaky.

In 1947, F. S. Bodenheimer of Hebrew University published an article on manna in *The Biblical Archaeologist*, laying out the case for . . . I kid you not . . . bug juice:

> [W]e find that manna production is a biological phenomenon of the dry deserts and steppes. The liquid honeydew excretion of a number of cicadas, plant lice, and scale insects speedily solidifies by rapid evaporation.

Bodenheimer explains his reasoning at some length:

"Man" is the common Arabic name for plant lice, and "man es-simma" (the manna of heaven) for honeydew. This confirms our views, because a number of small cicadas . . . are found in Sinai, southern Iraq, and Iran, and are locally called "man." They produce a product in small quantities which is similar to manna and which is used as a delicacy and as an ingredient in popular medicines. The large Cicada orni L. produces a like excretion on ash trees in Italy. The most famous manna product of the Middle East is the Kurdish manna which is collected by the thousands of kilograms every year in June and July. It is used for the preparation of special confections which are sold in the streets of Baghdad and elsewhere under the name of "man." This manna is also produced all over the general Kurdistan region in the extensive oak forest by a still undetermined aphid. This plant louse sucks on oak leaves and copiously excretes a honeydew which solidifies with fragments of the leaves. We were unfortunate in July 1943, when we visited one of the famous manna-producing forests near Penjween in northeastern Iraq, for the manna crop was a complete failure that year.

Israeli biologist Avionoam Danin had better luck when he set out into the wilderness in search of manna: "In July 1968 I found sweet white drops on Hammada [a kind of shrub that thrived in the area] at Wadi Feiran . . . It is not yet clear what caused the secretion, but near every drop was the skin of an insect."

So there you have it: manna, sweet stuff that squirts out of the rear end of bugs. Thanks be to the Lord.

The Israelites had better enjoy it, because this will be their daily ration for a long, long time. A whole generation will be born and raised on the stuff: "The Israelites ate manna forty years, until they came to a habitable land; they ate manna, until they came to the border of the land of Canaan." (Exodus 16:35, NRSV)

Let us pause here to note that manna makes another appearance later in the Bible, in the Book of Numbers, where it is clear that the attitude toward manna has changed. By now the Israelites are several years into their journey, and quite frankly a marinated monkey would be quite welcome at this point. As for manna, if they never see the stuff again it will be too soon:

[And] the Israelites also wept again, and said, "If only we had meat to eat! We remember the fish we used to eat in Egypt for nothing, the cucumbers, the melons, the leeks, the onions, and the garlic; but now our strength is dried up, and there is nothing at all but this manna to look at." (Numbers 11:4–8, NRSV)

Of course, man does not live by manna alone. The Israelites may be fixed for food, but they have to have water as well. There's no mention of it until they get to the border of Canaan, and then it takes a miracle to produce it. Moses pounds on a rock, and the water flows.

This is impressive, but what's really miraculous is how this million-plus flock of quarrelsome tribes managed to stumble around in the wilderness for forty years without a steady water supply. Maybe thirst prevented them from noticing that they were going in circles.

The Holy Land is not a big place. By any measure, the distance the tribes under Moses needed to travel if they followed a sensible route was no more than 420 kilometers, or about 250 miles. That's just a little more than the distance from Dallas to Houston, a four-hour drive, max. Granted the Israelites are walking, not driving. But even so, adults walk an average of 3 miles per hour, and they have the ultimate tour guide, the guy who made the world, God Almighty, peeping out of a cloud every now and then and muttering to Moses.

If they walked just eight hours a day, they'd cover the distance in about ten days. But let's remember, there are kids in that group, so we'll reduce the daily distance by half. They probably took the Sabbath off, so let's account for that. By giving them every benefit of the doubt, we can stretch a straight-line trek from Egypt to Canaan to 24 days. But the Bible tells us they did not get there for manna-filled decades.

The "wilderness of Sin," or Sinai Peninsula, is a dagger-shaped piece of land. It is about 150 miles (240 km) across at its widest and stretches roughly 240 miles (385 km) from north to south. If the Israelites zigged and zagged across the Sinai from top to bottom, the first crossing would have taken 18 days at most, including Sabbath stops. (A paltry 10 miles a day x 6 days = 60 miles per week.) Crossing back on the first diagonal might have been a bit longer; let's say 20 days. But then the pace picks up. Imagine these millions of Hebrews shuffling slantwise across the Sinai. Crossing on a diagonal would go a bit quicker each time, because we're starting to move down the narrowing peninsula. If each crossing put them just 10 miles further south, they would make a total of 24 trips before reaching the coral reefs at what is today Ras Muhammad National Park at the southern tip of the Sinai. If we allow generous rest stops, festivals, and Sabbaths, what's the longest that journey could have taken? By my reckoning, no more than nine months. At the end, with a million women and children in tow, could the Israelites have wheeled about to renew the journey northward? I can picture Moses standing on that spit of land with his back to the sea, shouting, "Okay, everybody, on the count of three . . . About face!"

Assuming they executed this million-man pirouette and zigzagged their way back up the Sinai, they would have found themselves back where they started in just 18 months. With the deity as their guide, how in the world did the Israelites manage to stay lost in the Sinai for 40 years? Did they go up and back more than two dozen times?

Archaeologists haven't found a trace of this mass migration. Not a sandal, not a midden, not a tentpole. Baptist minister and Christian apologist Rhett Totten has an explanation: he says that the Israelites were continually on the move, so they did not have time to make a lasting impression anywhere on the land. They "did not stay in any one place for more than a few days or perhaps a week or two," he writes, "and as a result, such brief encampments would not involve much land-clearing, and wouldn't leave significant amounts of objects behind which would survive the centuries." Then Rev. Totten goes on to claim something rather astounding: having endorsed the estimate of more than a million Hebrews setting out, he says, "[T]he judgment was that a whole generation would die in the wilderness, until there were only about 80,000 left at the end of 40 years, who finally entered into Canaan."

In other words, Totten is telling us that at least 920,000 Israelite males—more than the entire population of Austin, Texas—perished as they trekked 'round and 'round in circles, like Pooh and Piglet pursuing the woozles. In the desert climate of the Sinai, you might expect that archaeologists would stumble across the bones of nearly a million dead Israelites. But no, the Exodus left no trace. God's peeps move in mysterious ways.

11

 Joshua Fit the Battle of Jericho

"Joshua Fit the Battle of Jericho" is the title of one of my favorite spirituals. It conjures up one of the famous scenes in the Bible, when, in the words of that nineteenth-century "negro" song, "the walls came a-tumblin' down."

Now, unlike some locations named in the Bible, we know for sure that Jericho was a real place. Its remains lie on the western edge of the Jordan Valley, north of the Dead Sea. What's more, it has been excavated three times, and the last of these, conducted by Dr. Bryant G. Wood, research director of the Inerrantist Associates for Biblical Research, produced what Wood regards as strong evidence to support the biblical account of fallen walls.

But we're getting ahead of the story. The Book of Joshua opens scandalously: "After the death of Moses the servant of the Lord, the Lord spoke to Joshua son of Nun." Well, now! Who even knew there were nuns back then?

However humble Joshua's background may be, God has marching orders for him:

> Now proceed to cross the Jordan, you and all this people, into the land
> that I am giving to them, to the Israelites. Every place that the sole of
> your foot will tread upon I have given to you, as I promised to Moses.
> From the wilderness and the Lebanon as far as the great river, the river
> Euphrates, all the land of the Hittites, to the Great Sea in the west
> shall be your territory. No one shall be able to stand against you all the

days of your life. As I was with Moses, so I will be with you; I will not fail you or forsake you. Be strong and courageous; for you shall put this people in possession of the land that I swore to their ancestors to give them. Only be strong and very courageous, being careful to act in accordance with all the law that my servant Moses commanded you; do not turn from it to the right hand or to the left, so that you may be successful wherever you go. (Joshua 1:2–8, NRSV)

Yes, once again, God is planning genocide, and this time on a scale not seen since he drowned every creature except those aboard the ark. On the chopping block are: the Canaanites, Hittites, Hivites, Perizzites, Girgashites, Amorites, and Jebusites. Their only crime, apart from adopting silly names, is being in the wrong place at the wrong time. They dwell on land that God has promised to Moses and his people.

To achieve his foreign-policy aims, all God needs is for the Israelites to obey orders without question. For the rest, the Lord arranges a miraculous stoppage in the flow of the river Jordan, a miraculous collapse of the walls of Jericho, and a little help from a traitorous prostitute.

The lady in question is one Rahab of Jericho, who conceals a pair of spies that Joshua sends to get the lay of the land. She hides them on her roof and sends the king's guard on a false pursuit. After they've left, she explains her betrayal:

I know that the Lord has given you the land, and that dread of you has fallen on us, and that all the inhabitants of the land melt in fear before you. For we have heard how the Lord dried up the water of the Red Sea before you when you came out of Egypt, and what you did to the two kings of the Amorites that were beyond the Jordan, to Sihon and Og, whom you utterly destroyed. As soon as we heard it, our hearts melted, and there was no courage left in any of us because of you. The Lord your God is indeed God in heaven above and on earth below. Now then, since I have dealt kindly with you, swear to me by the Lord that you in turn will deal kindly with my family. Give me a sign of good faith that you will spare my father and mother, my brothers and sisters, and all who belong to them, and deliver our lives from death. (Joshua 2:8–13, NRSV)

So, she sells out Jericho in return for her family's safe passage. Some ethics. But, there's no opportunity to send Rahab to rehab, so the story rolls on.

The spies return safely, and the invasion gets under way. Joshua orders his priests to bring out the Ark of the Covenant—a decorated chest containing texts of God's contract with the Israelites—and to walk ahead of the army as far as the waters of the Jordan. They do, and as soon as their feet touch the water, "the waters flowing from above stood still, rising up in a single heap far off at Adam, the city that is beside Zarethan, while those flowing toward the sea of the Arabah, the Dead Sea, were wholly cut off. Then the people crossed over opposite Jericho." (Joshua 3:16, NRSV)

Now, for this to happen just as written would be fortuitous, but a miracle is not required for the river to temporarily halt its flow. Instead, it needs precisely the same natural phenomenon as can explain the subsequent collapse of the walls: an earthquake. Is such a thing plausible in the region? Oh, quite.

Jericho has the misfortune to sit along a major fault. Indeed, the nearby Dead Sea is a low point in a vast rift that is slowly tearing the Middle East in twain. A 1927 earthquake centered near Jericho prompted geologists to investigate, and they found evidence of a long history of temblors in the same location. What's more, there's evidence that the earthquakes often cause a temporary damming of the Jordan River.

Could this have happened at just the right time for Joshua to lead an invasion across the Jordan? Sure. Why not? Here's the biblical set up for what happens next:

> The Lord said to Joshua, "See, I have handed Jericho over to you, along with its king and soldiers. You shall march around the city, all the warriors circling the city once. Thus you shall do for six days, with seven priests bearing seven trumpets of rams' horns before the ark. On the seventh day you shall march around the city seven times, the priests blowing the trumpets. When they make a long blast with the ram's horn, as soon as you hear the sound of the trumpet, then all the people shall shout with a great shout; and the wall of the city will fall down flat, and all the people shall charge straight ahead." (Joshua 6:2–5, NRSV)

Apart from the falling walls, what's described here amounts to pretty routine siege tactics. If you are the besieger, you don't want to wait for starvation to set in. You want to terrify your enemy into surrender. An army on parade has precisely that effect—especially if it is accompanied by loud martial music. Just ask Napoleon, Hitler, or any of your other leading dictators. Note that Joshua's army was clearly bent on depriving the inhabitants of sleep until one day . . .

> Joshua rose early in the morning, and the priests took up the ark of the Lord. The seven priests carrying the seven trumpets of rams' horns before the ark of the Lord passed on, blowing the trumpets continually. The armed men went before them, and the rear guard came after the ark of the Lord, while the trumpets blew continually. . . . They did this for six days.
>
> On the seventh day they rose early, at dawn, and marched around the city in the same manner seven times. It was only on that day that they marched around the city seven times. And at the seventh time, when the priests had blown the trumpets, Joshua said to the people, "Shout! For the Lord has given you the city. The city and all that is in it shall be devoted to the Lord for destruction. Only Rahab the prostitute and all who are with her in her house shall live because she hid the messengers we sent . . .
>
> As soon as the people heard the sound of the trumpets, they raised a great shout, and the wall fell down flat; so the people charged straight ahead into the city and captured it. Then they devoted to destruction by the edge of the sword all in the city, both men and women, young and old, oxen, sheep, and donkeys. (Joshua 6:13–21, NRSV)

Genocide accomplished!

Could shouting have brought down the walls? Highly unlikely. Mud walls are not crystalline, so they would not have a single resonant frequency. What's more the energy generated by a large group is not all that impressive. The roar of a football stadium may be stunning to the ears, but it never brings down the house. What could have done the job, however, is an aftershock. A primary temblor powerful enough to block the Jordan River would almost certainly produce aftershocks. One of those could topple the already weakened mud walls of Jericho. This is what Dr. Wood found: evidence of mud walls that had collapsed. He's convinced that this validates the biblical account of the battle of Jericho. Most other biblical scholars disagree, but we'll come back to that in a moment.

Once the slaughter is complete, Joshua puts a curse on the remains of Jericho and the Israelites move on. But trouble soon follows. When the Israelites attack the city of Ai, they are routed. How can this be, with God on their side? It's the usual problem: sin.

Joshua soon roots out the source: in his ranks one Achan has kept some of the booty instead of turning it over to God's treasury (conveniently held in trust by the priests). The Israelites resolve the issue by stoning Achan and all his children and even his animals to death. The Lord exhales an "ahhhh" of satisfaction and gives them a battle plan that provides for the slaughter of every man, woman, and child in Ai.

This may be horrifying, but it's entirely credible. The stoppage of the river and the tumbling of the walls are explicable by natural means. Genocide continues even in our time, and there are good reasons to believe it was much more common in the past.

But there's one whopper that won't stand: the halting of Earth. Here's what we're told. All the Amorite kings gang up on a servile ally of the Israelites, causing Joshua to bring his army to the rescue. An Israelite rout ensues. Joshua, needing more time to slake his bloodlust, calls out: "Sun, stand still at Gibeon, and Moon, in the valley of Aijalon." And then, the Bible says, "the sun stood still, and the moon stopped, until the nation took vengeance on their enemies." (Joshua 10:13, NRSV)

The account goes on to add important details: "The sun stopped in midheaven, and did not hurry to set for about a whole day. There has been no day like it before or since."

Can this have actually happened? Many apologists have strained at explanations. Some have invoked an eclipse, but this would require the moon to pass in front of the sun—not stand still—and would have been over in less than eight minutes.

Some have wondered if the sun could have hovered over the battlefield while Earth continued to turn. In principle, it is possible that the entire universe rotated around Earth for 24 hours—there are no privileged frames of reference in physics. But this would require most of the objects in the universe to travel faster than the speed of light relative to Earth for a 24–hour period, and that's a physical impossibility.

Equally impossible would be moving the sun around Earth for a day—the solar system would be thrown into catastrophic disarray. Imagine what it would be like for a Martian, sitting on his porch, sipping green tea, when in response to a prayer from some bloodthirsty Israelite the sun suddenly takes off at 24.5 million miles an hour. That's how fast it will have go to keep up with the rotating Earth. To accelerate to 3.5 percent of the speed of light from a standing start would cause a complete deformation of the sun. If pushed with equal force at every point it would turn into something like a giant tulip. Just as when a stoplight changes to green, a truck will accelerate much more slowly than the cars beside it, so the Sun's massive core would trail behind its outer layers. If the force were concentrated on the core, then in the dying moments of his life, the Martian would see something like a huge, blazing comet streaking away, doing all 5 billion years' worth of remaining fusion in one gallant burst. Then, all hell would break loose as the planetary orbits went into disarray.

So, looking to other solutions to the Joshua problem, some argue that Earth merely slowed down, making for a long day. This explanation has numerous problems.

When an airplane lands, it is merely slowing down, but there's a reason they make you fasten your seatbelt. Any object that's left loose will, as the flight attendants say, tend to shift. Earth's rotational speed at the equator is about a thousand miles an hour, more than six times the speed of an airliner at landing. If you were to cut that speed by half, even over a matter of hours, the oceans and every other body of water would spill over the land in a flood that would put Noah's to shame.

It would also likely make the moon spin. The moon is tidally locked to Earth, so its face never moves. But cut the rotational energy of Earth and by the laws of conservation it would have to go somewhere. My hunch is that the moon would commence to pirouette.

If it didn't melt, that is. How do you slow a body as massive as Earth? Even the hand of God would not be gentle enough to prevent intense friction, as every molecule on the planet slammed into its neighbor. The heat storm would be catastrophic.

And finally, how do you restart it? After slowing Earth, you'd have to add back an equal amount of rotational momentum to speed it up again. Perhaps two streaking chunks of neutron star, coming from opposite directions and simultaneously clipping Earth at, say, equatorial Africa and Brazil, might do it, but after the resultant thermonuclear devastation and its aftermath there'd be no one left to tell the tale.

As for Earth literally stopping, it's a nonstarter. Every problem listed above would be increased exponentially. Which brings us back to the mud walls. While it's entirely possible that Dr. Wood is right about an earthquake felling the walls of Jericho, most biblical scholars doubt that there's any relation to the conquest of the region. Other evidence clearly shows that the Israelites did not overrun the Promised Land in a blitzkrieg. It's not even clear that they invaded at all.

That there was routine violence is beyond dispute. But, given the absence of support for the claim of a mass exodus from Egypt or a 40-year sojourn in the Sinai wilderness (see previous chapter), the case against an invasion seems strong. But Joshua may have had his day.

According to the Bible, the blood-lusting general led his army to a climactic battle at Hazor, an opulent city whose remains are still being excavated. In 2012, archaeologists we earlier mentioned, Amnon Ben-Tor and Sharon Zuckerman, led a team that discovered clear evidence of a catastrophic fire that destroyed the main palace at Hazor. However, they split over how to interpret the evidence. Ben-Tor saw it as clinching the case for Joshua's army; Zuckerman thought it supported the idea of a local uprising. The jury is still out.

In any event, the one point of broad consensus among archaeologists is that the ancient state of Israel emerged gradually. If Yahweh was leading the battle, he took his sweet time. The last words go to biblical scholar Bart Ehrman:

> There are no references in any other ancient source to a massive destruction of the cities of Canaan. Archaeologists have discovered that few of the places mentioned were walled towns at the time. Many of the specific cities cited as places of conquest apparently did not even exist as cities at the time. This includes, most notably, Jericho, which was not inhabited in the late 13th century BCE, as

archaeologists have decisively shown. The same thing applies to Ai and Heshbon. These cities were neither occupied, nor conquered, nor re-inhabited in the days of Joshua. . . . There are, to be sure, some indications that some towns in Canaan were destroyed at about that time (two of the twenty places mentioned as being destroyed by Joshua were wiped out at about the right time: Hazor and Bethel). But that is true of virtually every time in antiquity: occasionally towns were destroyed by other towns or burned or otherwise abandoned. We are left, then, with a very big problem. The accounts in Joshua appear to be non-historical in many respects. This creates a dilemma for historians, since two things are perfectly clear: (a) eventually there was a nation Israel living in the land of Canaan; but (b) there is no evidence that it got there by entering in from the East and destroying all the major cities in a series of violent military campaigns. Where then did Israel come from?

Ay, caramba, Bart. Good question.

12

 The Numbers Game

After the excitement, drama, and slaughter of Exodus, the Bible settles down into some downright boring books. They are, essentially, the IRS Code of the Israelites, instructing ancient people in how to follow the law and pony up. But miracles keep dribbling in.

The Book of Numbers opens with God telling Moses to conduct a census of the male Israelites. It turns out to be a draft registration—and guess who gets exempted from going to war? The extended family that produces priests, known to all as the Levites.

In Numbers, God supposedly sets down rules that give that priestly clan a sweet deal. What's miraculous, if you like, is that the priests managed to convince so many people that God, supposedly the creator of everything, needs chunks of it back as bribes. When a business opens up, it usually gives away coupons, door prizes, and free samples. But at Yahweh's grand openings, it's the other way 'round. Just look at the list of "offerings" brought by the tribal leaders for the unveiling of a new altar.

> This was the dedication offering for the altar, at the time when it was anointed, from the leaders of Israel: twelve silver plates, twelve silver basins, twelve golden dishes, each silver plate weighing one hundred thirty shekels and each basin seventy, all the silver of the vessels two thousand four hundred shekels according to the shekel of the sanctuary, the twelve golden dishes, full of incense, weighing ten shekels apiece according to the shekel of the sanctuary, all the gold

of the dishes being one hundred twenty shekels; all the livestock for the burnt offering twelve bulls, twelve rams, twelve male lambs a year old, with their grain offering; and twelve male goats for a sin offering; and all the livestock for the sacrifice of well-being twenty-four bulls, the rams sixty, the male goats sixty, the male lambs a year old sixty. (Numbers 7:84–88, NRSV)

It's a freakin' menagerie, weighed down with silver and gold. Where is God gonna keep all this stuff? The Bible provides an answer: "Among all the sacred donations of the Israelites, every gift that they bring to the priest shall be his."

Wow. If you had to live in ancient Israel, a Levite is who you wanted to be. No army service, no tilling the soil or herding cattle, and a world-class retirement plan.

The Lord spoke to Moses, saying: This applies to the Levites: from twenty-five years old and upward they shall begin to do duty in the service of the tent of meeting; and from the age of fifty years they shall retire from the duty of the service and serve no more. They may assist their brothers in the tent of meeting in carrying out their duties, but they shall perform no service. (Numbers 8:24–26, NRSV)

What a deal. Retired at fifty, and all the smoked meat you can eat. It's just too bad that golf hadn't yet been invented. Then, truly, the Levites would have enjoyed heaven on Earth.

They say the age of miracles ended with the apostles, but this one sure hasn't. In modern-day Israel, the ultra-Orthodox, or Haredi, have emerged as free riders on the state. In a country perpetually at war with neighboring states and a large chunk of its own occupied population, the Haredi have long refused military service. Their rabbis argue that military service exposes their youth to "adultery, homosexuality . . . and all the rest," and that the nation is better off obeying God's laws than defending itself. Yet, like the Levites before them, they don't actually seem to mind when fellow Israelis put their lives on the line to defend the nation.

Until 2014, they were exempted *en masse* from military duty on the grounds that they were studying religion, but finally an exasperated public pressed the Knesset to pass a new law requiring young Haredi to do either military or civil service. They have generally resisted compliance. What's more, a majority of Haredi adults in Israel decline employment, choosing to scrape by on public assistance and charity. Looking back to the opening of Numbers, this may qualify as the longest-lasting miracle of them all.

Can science explain this? Actually, it can. Consider the case for the Levites as parasites.

Nature is full of bloodsuckers. From an evolutionary standpoint, this is an inevitable development. All life requires energy to function, and useful energy is hard to come by. Plants patiently extract it from the sun. Herbivores graze hour after hour, patiently harvesting it from plants. Carnivores stalk their prey and then engage in an exhausting hunt to the kill. But parasites? They make use of shortcuts, scams, and diversions to steal the energy that others have gathered.

The cuckoo is one of many brood parasites. It lays its eggs in the nests of other birds, and then its egg hatches first and pushes the other eggs out of the nest. It then fools the parents into feeding it, protecting it, and maybe even paying for college. Pretty mean trick, eh?

Wasps in the genus *Glyptapanteles* are even meaner. They inject their eggs into living caterpillars, who are then consumed from within. Parasites don't have it all their own way, though. Hosts evolve various strategies to resist. Trees develop bark, plants produce toxins, and apes groom one another to pick off bloodsuckers. And people? We humans have an instinctive disgust reaction to parasites that threaten us. Food infested with worms repels people everywhere. We have no such reaction to parasites that affect distantly related species. For example, mistletoe is a parasite on trees, but we consider it a mildly romantic decoration. Yet, you never see any couple kissing under a tapeworm.

This instinctive reaction carries over into human social relations. It angers us to see someone taking it easy while others work hard. That's why the Bible condemns it as a sin, and why the Roman poet Ovid wrote, "sloth wastes the sluggish body, as water is corrupted unless it moves." As writers often do, he intuitively connects sloth with our *instinctive* disgust response to stagnant water, teeming with parasites.

This element of evolutionary psychology is frighteningly potent. By branding Jews as parasites, Hitler was able to stir the German people to blind rage against them. Ronald Reagan, an outsider even in his own party, was able to cruise victory over a sitting president in part because of his skillful use of the myth of a Cadillac-driving, implicitly black welfare queen.

Still, to live as a parasite must be a great temptation. How can it be done without evoking deadly rage in others? One way is to be born into a rich and powerful family, like the Kim clan in North Korea. But even then, you must constantly guard against an uprising of the peasants, traitors in your coterie, or a coup by the palace guard.

A better solution is to claim the imprimatur of God. If your all-powerful, invisible yet omnipresent pal in the sky says, "Among all the sacred donations of the Israelites, every gift that they bring to the priest shall be his"—well, who's gonna argue?

Of course, maintaining the special status that exempts you from working in the fields or fighting in the ranks requires some upkeep. You have to perform rituals, make a show of virtue, and set yourself apart from the masses. But for most of history, in most societies, it's been a bargain worth the price.

If you're a religious person, this view of priests may come as an affront. But set aside your own religion for a moment and consider another. Let's assume for the moment that there are no Japanese readers of this book. (Ahem. Note to publisher: let's change that for the Japanese edition!) Shinto is the native religion of Japan, complete with rituals, shrines, and a priesthood. The job of the priest is to maintain proper relations between a human community and the *kami*, or gods. Shinto is animistic, so the *kami* are thought to be invisible spirits that inhabit trees, rocks, and other features of nature. To ensure health, success, and other desired outcomes, the priest dons beautiful robes and conducts rituals, for which tribute is paid.

Now, presumably you and I don't believe there are *kami* hovering in trees, rocks, or ponds. If so, the priest is a pious fraud. For at least 100 generations, he has accepted rice, fish, sake, or cash from hard-working villagers for an act that has no power to improve their lives. But let's be clear: he may well believe in his powers. In fact, he's much more likely to succeed in the pretense if he sincerely believes that his incantation, his sprinkling of ritual sake, or a young girl's performance of a sacred dance will bring a good rice harvest. Nevertheless, he is a parasite all the same. His bloodsucking on the labor of others bypasses the instinctive outrage that most parasites trigger because he's able to maintain the illusion of providing a real, rather than an illusory, service.

Or is it? Some would argue that, whether or not the gods are real, the function of religion provides a real return to the believers in the form of solidarity, solace, and confidence. The evolutionary anthropologist David Sloan Wilson makes this argument, and offers some evidence for it. (We'll return to his claims in the final chapter.) Of course, assuming that's true, there could be parasitical versions that cynically mooch off the psychological needs of their flocks. Televangelists come to mind.

Still, if a Shinto priest provides value for money, it may be true for a Levite as well, along with every other "authentic" priest, imam, shaman, and what have you in history. So there are, perhaps, two scientific explanations for the miracle by which God made the tribal leaders meekly hand over gold dishes, silver trays, bulls, rams, and goats by the dozen to a bunch of indolent priests who neither tilled the field nor wielded spear or sword. One is that priests are parasites; the other is that they inspire a self-sacrificing social cohesion that makes the whole tribe prosper in relation to other tribes that haven't learned this useful mass delusion. What's more both explanations may be true!

All of this is just a warm-up for one of the most amusing, confusing, self-bollixing miracles in all of Scripture. We speak, of course, of Balaam's ass (meaning his ride, not his rear).

Balaam was a kind of witch doctor or wizard-for-hire. If you needed someone cursed anywhere on the plains of Moab then Balaam was your go-to guy. So, when the Israelites came spilling out of the Sinai, all million-plus of them, the Moabites reacted like Republicans facing a boatload of refugees:

> Moab was overcome with fear of the people of Israel. And Moab said to the elders of Midian, "This horde will now lick up all that is around us, as an ox licks up the grass of the field." Now Balak son of Zippor was king of Moab at that time. He sent messengers to Balaam . . . to

summon him, saying, "A people has come out of Egypt; they have spread over the face of the earth, and they have settled next to me. Come now, curse this people for me, since they are stronger than I; perhaps I shall be able to defeat them and drive them from the land. . . . So the elders of Moab and the elders of Midian departed with the fees for divination in their hand; and they came to Balaam, and gave him Balak's message. (Numbers 22:3–7, NRSV)

This being the Middle East, a long haggle begins. Balaam plays a curious card in the negotiation: he says God told him not to go—and sure enough, that's who's holding things up. But the Moabites keep raising the ante until God gives in:

That night God came to Balaam and said to him, "If the men have come to summon you, get up and go with them; but do only what I tell you to do." So Balaam got up in the morning, saddled his donkey, and went with the officials of Moab. (Numbers 22:20–21, NRSV)

But apparently this is one of those grudging, "Fine, then. Go if you want! What do I care?" decisions, for the Bible tells us: "God's anger was kindled because he was going, and the angel of the Lord took his stand in the road as his adversary." And that's when things get really weird. There's this angel of the Lord, sword in hand, blocking the path . . . but only the donkey can see it:

Then the angel of the Lord stood in a narrow path between the vineyards, with a wall on either side. When the donkey saw the angel of the Lord, it scraped against the wall, and scraped Balaam's foot against the wall; so he struck it again. Then the angel of the Lord went ahead, and stood in a narrow place, where there was no way to turn either to the right or to the left. When the donkey saw the angel of the Lord, it lay down under Balaam; and Balaam's anger was kindled, and he struck the donkey with his staff.(Numbers 22:24–27, NRSV)

What is the point, you may wonder, of an invisible angel? It's kind of like advertising on a Times Square megascreen that's all black, all the time. Well, not to worry. God has a plan: he'll make the donkey talk.

Then the Lord opened the mouth of the donkey, and it said to Balaam, "What have I done to you, that you have struck me these three times?" Balaam said to the donkey, "Because you have made a fool of me! I wish I had a sword in my hand! I would kill you right now!" But the donkey said to Balaam, "Am I not your donkey, which you have ridden all your life to this day? Have I been in the habit of treating you this way?" (Numbers 22:28–30, NRSV)

Now, making a donkey talk is a pretty good trick. Many years would pass until it was repeated, this time by the Dreamworks animators who created *Shrek* and hired Eddie Murphy to voice The Donkey. But, without a voice actor, could it be done?

Even our closest relations, the great apes, are physically and mentally incapable of speech. The human larynx, the part of the throat where speech begins, is exceptional. According to the Max Planck Institute for Psycholinguistics, "The organs within the vocal tract, such as larynx muscles and vocal cords, cannot be moved as freely and coordinated as in humans, especially not at a comparable speed. For this reason, we cannot talk with apes."

Now, apes can communicate—a damn sight more than donkeys can. They have a suite of grunts, snuffles, and gestures that allow them to establish hierarchy, coordinate attacks on others, and seek help. But even the most sophisticated of primate communicators lacks the intellectual capacity (not to mention the Jewish nuance) to come up with a plaintive statement like this: "Am I not your donkey, which you have ridden all your life to this day? Have I been in the habit of treating you this way?" You can almost hear the donkey snort and then add, "Vat am I, chopped liver?"

For a realistic contrast, take Koko the gorilla. Perhaps the most successful student of sign language among the great apes, Koko by the age of five had mastered a little over 500 signs. That's impressive, but a human child will typically have a vocabulary seven to ten times that large by the same age. What's more, Koko had few abstract concepts, and certainly not the idea of a lifetime. She would have been utterly incapable of pondering whether she had habitually provided excellent transport services to her keepers.

This presents a double-barreled problem for any deity looking to use natural means to make an ass talk. First, how do you fine tune the vocal cords of an animal best known for braying (hee-haw!) to make it into an articulate interlocutor? Second, how do you reprogram its brain?

The vocal cords of a donkey are quite different from those in humans. Still, it is conceivable that they could be manipulated to form something like speech. After all, a kazoo can make sounds that are midway between human language and a donkey's bray. With some expert tweaking maybe the bray could sound like a brogue.

The real problem lies in the brain. Human speech is orchestrated by a little clump of brain tissue called Broca's area. Actually, "conducted" may be the better word. Like sections of the orchestra, many areas of the brain participate in the production of speech, but it appears that Broca's area actually directs the vocal cord, tongue, and mouth movements that result in talk.

French physician Paul Broca discovered this following the death of one of his patients nicknamed "Tan," after the only sound the poor man could produce. Broca met Tan in 1861, decades after his ability to speak had been impaired. By that time, the patient's condition had greatly deteriorated. Much of his body was paralyzed, his vision was failing, and his spirits, naturally enough, were deeply depressed. Broca could not heal Tan, but he did find a way to communicate with him, using gestures Tan could make with his left hand. Before Tan died, some five months later, Broca had proof that Tan could think just fine; he simply could not speak.

After Tan's death, an autopsy revealed a lesion in the left frontal region of his brain. When an autopsy on a second patient similarly afflicted showed similar damage to the brain, the case was clinched.

Broca's area may not be completely unique to humans—controversial research suggests that other apes to some degree share it—but beyond all doubt it has evolved unique versatility in humans. However much training an ape may have, its vocal skills never rise above those of its anthropoidal peer group. That's why researchers resort to sign language and symbol manipulation to talk with gorillas and other simian subjects.

For sure, donkeys don't come anywhere close to having the communicative skill of a baboon, let alone that of a human. And even if we somehow grant Balaam's ass a freak transformation of its wetware, there remains a software bug. How in the world did a donkey learn ancient Hebrew?

Dogs, more than any other animal, have evolved skills to attune themselves to human behavior. They interpret our facial expressions, they get our gestures, and they catch on to our commands. But even the smartest dogs' comprehension of human language tops out at about 200 words. Donkeys? Not so much. I hate to stereotype, but there's a reason behind the expression "dumb ass."

So, short of magic, is there any possible way that the Balaam story could be true? Put a little differently, is there any way Balaam could have been convinced that his donkey talked? Sure there is. God would just have had to make use of the same technology that a modern-day prankster would use: a Bluetooth speaker.

As for the angel in the road, well, a little dry ice and an LED projector would do the job. God would just have had to time-travel to the twenty-first century, borrow some battery-powered equipment, and zip back to the ancient Holy Land. Miracle accomplished.

That's cheating, you say? Not at all. It's technomancy. If we just leap ahead a few thousand years, it's evident that such illusions can be convincingly created by natural means. Clarke's Third Law comes to mind: "Any sufficiently advanced technology is indistinguishable from magic."

The great science fiction writer may have been implying a corollary: to those who are ignorant of science and technology, the extraordinary must appear to be magic. That would certainly describe Balaam and, indeed, all the writers of the Bible. They were intuitively aware of the regularities of nature, just as we are, but they had no notion that scientific explanations could account for many if not all of life's surprising turns.

But there's more to the story of Balaam's ass. If the name sounds faintly funny to our ears, the story likely provoked guffaws back in the day of its telling. Humor doesn't hold up well across the ages, but many scholars, including some conservatives, view this tale as a satire never intended to be taken literally. Writing in *Tektonics*, conservative apologist James P. Holding embraces this view:

I have shown here many times that it is a mistake to read the Bible through a modern lens, and in so doing define "inerrancy" in a modern fashion. The 3400 years that separate us make a difference.

Closer to the truth, I think, is the solution proposed by [Schubert] Spero. It fits in nicely with the archaeological data that seems to have confirmed that Balaam actually existed as a pagan diviner and prophet.

What he proposes is that in this story of Balaam, we have a quite intentional contradiction made for the purpose of poking fun at a very popular diviner who was no match for the true God. In other words, the narrative is interrupted for a few moments of satire—and it's only because we are reading through a modern lens that we don't realize this.

So, in the end there may be no miracle to explain here. But the discovery of a satire tucked away in the Bible raises some interesting possibilities.

Adam's rib? You gotta be joking.

13

After Joshua's romp through Canaan, the Bible settles down to a bunch of second-rate miracles. In Kings, the prophet Elijah prays and has it both ways—first drought, then rain. A miracle? You be the judge.

His protégé, Elisha, pulls off what has to be the most petty, mean-spirited, and vile "miracle" in all of Scripture. If you read this book's introduction, you know what's coming. On his way to the city of Bethel, Elisha runs into an unruly group of boys. Now, this is not a gang. They do not rob, or beat, or molest him. They do not so much as touch a hair on his head—although in fairness, they could not have done so even if they'd wanted to. Let's let the Bible tell the tale:

> [Elisha] went up from there to Bethel; and while he was going up on the way, some small boys came out of the city and jeered at him, saying, "Go away, baldhead! Go away, baldhead!" (2 Kings 2:24, NRSV)

Got that? They jeered. They mocked. They called him "baldy." So, how does Elisha respond?

> When he turned around and saw them, he cursed them in the name of the Lord. Then two she-bears came out of the woods and mauled forty-two of the boys. From there he went on to Mount Carmel, and then returned to Samaria. (2 Kings 2:25, NRSV)

Is this a miracle? More of an improbability, really. As I noted in the introduction, bears are native to the region. If only the children had been attacked by Galapagos tortoises or a herd of exceptionally fierce capybaras. Seeing as those animals live only in South America, that would have been closer to a miracle. But to be on the safe side, kids, if you run into a prophet of the Lord, whatever you do, don't mention his follicular shortcomings.

Several other Old Testament miracles in this book are worth mentioning because they foreshadow—not to say, undercut—those yet to be performed in the New Testament as proof that a certain someone is the Son of You-Know-Who.

For instance, in 2 Kings 4, Elisha resurrects a boy from the dead. The backstory is somewhat amusing, so let's have a look before we plunge into the miracle.

> One day Elisha was passing through Shunem, where a wealthy woman lived, who urged him to have a meal. So whenever he passed that way, he would stop there for a meal. She said to her husband, "Look, I am sure that this man who regularly passes our way is a holy man of God. Let us make a small roof chamber with walls, and put there for him a bed, a table, a chair, and a lamp, so that he can stay there whenever he comes to us." (2 Kings 4:8–10, NRSV)

Elisha is grateful and asks what he can do for her. The woman demurs, suggesting she has all she needs. Elisha then asks his servant: "What then may be done for her?" Gehazi answered, "Well, she has no son, and her husband is old."

Hmmm . . . Interesting. Seems that Elisha provides fertility services on the side. Just how remains a matter of speculation.

> When he had called her, she stood at the door. He said, "At this season, in due time, you shall embrace a son." She replied, "No, my lord, O man of God; do not deceive your servant." The woman conceived and bore a son at that season, in due time, as Elisha had declared to her. (2 Kings 4:15–17, NRSV)

Happy ending, right? But trouble lies ahead.

When the child was older, he went out one day to his father among the reapers. He complained to his father, "Oh, my head, my head!" The father said to his servant, "Carry him to his mother." He carried him and brought him to his mother; the child sat on her lap until noon, and he died. She went up and laid him on the bed of the man of God, closed the door on him, and left. (2 Kings 4:18–21, NRSV)

The Shunammite woman, who like most women in the Bible seems to lack a name, rushes off to find Elisha. She throws herself at his feet and begs him for help. But Elisha's a busy prophet. At first, he's only willing to send his staff—both figuratively and literally. "He said to Gehazi, 'Gird up your loins, and take my staff in your hand, and go. If you meet anyone, give no greeting, and if anyone greets you, do not answer; and lay my staff on the face of the child.'"

But the mother refuses to settle for the servant, and finally the prophet relents and ambles on down to have a look for himself. Good thing, because in the meantime Gehazi comes back and reports that the staff has done the boy no good.

When Elisha came into the house, he saw the child lying dead on his bed. So he went in and closed the door on the two of them, and prayed to the Lord. Then he got up on the bed and lay upon the child, putting his mouth upon his mouth, his eyes upon his eyes, and his hands upon his hands; and while he lay bent over him, the flesh of the child became warm. (2 Kings 4:32–34, NRSV)

Okay, that's more than a little creepy, but sometimes to effect a cure one must be cruel to be kind. Maybe this is a kind of biblical CPR. Sure enough, the mouth-to-mouth treatment works.

He got down, walked once to and fro in the room, then got up again and bent over him; the child sneezed seven times, and the child opened his eyes. Elisha summoned Gehazi and said, "Call the Shunammite woman." So he called her. When she came to him, he said, "Take your son." She came and fell at his feet, bowing to the ground; then she took her son and left. (2 Kings 4:35–37, NRSV)

It's a miracle! Or is it? An alternative explanation is close at hand. Until Elisha did his mouth-to-mouth, full-body-press thing, the poor Shunammite kid may have been in a coma. What could have brought that on? How 'bout encephalitis, a swelling of the brain? And how might that have happened? Through infection by an encephalitic virus, such as the West Nile virus, brought to you by your friendly local mosquitoes.

The meager medical description that we have is consistent with this diagnosis: the severe headache, the listlessness, and then the loss of consciousness. You might think that anyone would know the difference between death and a coma, but that's only because most of us never deal with either face-to-face. We leave that to the experts.

Even they sometimes find it difficult to discern the line between life and death. For the ancients, it was a mess. Thousands of years after this tale was told, Moses Maimonides, a wise Jew living in the twelfth-century Islamic empire, devised a method to test whether patients were dead or in a coma. He recommended placing a piece of glass under the nostrils of the patient to see if it fogged up, or alternatively a feather to see if any faint breath would ruffle it. Even when these tests produced no sign, Maimonides recommended waiting another hour before concluding the patient was dead.

No wonder, then, that Ma Shunammite refused to give up hope. But what of the cure? Well, the first thing we must keep in mind is that, fatalities aside, most diseases go away. The body cures itself. With West Nile virus, children are more likely than adults to suffer a coma, but they are also more likely to recover.

And what is more likely to wake a child from a coma than to have some lecherous old prophet doing a full-lip face-plant on him? If there are any convincing resurrection tales in the Bible, this one does not make the grade. The only miracle here is that on waking up the kid didn't blurt out, "Ma, who's this bald weirdo on top of me?"

Cue the bears.

14

 Jonah and the Great Fish

Just about everyone knows the story of Jonah and the whale, but very few know why Brother J wound up in there. It's a pretty interesting tale.

The Lord is unhappy about vice in Nineveh. Apparently, it's the Las Vegas of its day—there's a whole heap of fornicating, gambling, and whatnot going on. In the good old days, the King of Kings would have leaned over, tossed a little fire, a little brimstone, and poof! But God's gone all soft and liberal by this point, so he just commands Jonah to get on up there and preach.

Understandably, Jonah is none too thrilled with the assignment. Even today Nineveh is not a choice destination. Let's put it this way: it lies along the Euphrates by what is now Mosul—a city in northern Iraq lately ruined by ISIS. Back then, Nineveh was home to the goddess Ishtar and her followers. And remember, Jonah is Jewish. So can we really blame him when he turns tail? Well, someone does.

> Jonah set out to flee to Tarshish from the presence of the Lord. He went down to Joppa and found a ship going to Tarshish; so he paid his fare and went on board, to go with them to Tarshish, away from the presence of the Lord.
>
> But the Lord hurled a great wind upon the sea, and such a mighty storm came upon the sea that the ship threatened to break up. Then the mariners were afraid, and each cried to his god. (Jonah 1:3–5, NRSV)

This is one of numerous spots where the Bible casually acknowledges the existence of other gods, but in this instance none of them proves reliable. So, they resort to a more practical solution. They cast lots. Yes, apparently the way to find out who is concealing a secret is to draw straws. Jonah comes up short, and they pounce. Here's where it gets even weirder.

> Then they said to him, "Tell us why this calamity has come upon us. What is your occupation? Where do you come from? What is your country? And of what people are you?" "I am a Hebrew," he replied. "I worship the Lord, the God of heaven, who made the sea and the dry land." Then the men were even more afraid, and said to him, "What is this that you have done!" For the men knew that he was fleeing from the presence of the Lord, because he had told them so. (Jonah 1:8–10, NRSV)

They knew he was fleeing God, but did it not occur to anyone that this might be the reason for the tempest? Well, at least now that they know they'll do the sensible thing. Even Jonah realizes that.

> "Pick me up and throw me into the sea; then the sea will quiet down for you; for I know it is because of me that this great storm has come upon you." Nevertheless the men rowed hard to bring the ship back to land, but they could not, for the sea grew more and more stormy against them. (Jonah 1:12–13, NRSV)

So, giving in to the inevitable, they chuck Jonah overboard, and *alakazam*! Calm seas and fair weather all around. For Jonah, though, things take a turn for the worse:

"Now the Lord provided a huge fish to swallow Jonah, and Jonah was in the belly of the fish three days and three nights."

This one line has caused a huge controversy in modern times. Smarty-pants atheists have pointed out that a whale is a mammal not a fish. Mammals are milk-producing, air-breathing animals that give birth to live young. Fish breathe water and lay eggs.

To which I, a not-so-smart atheist, point out that the text doesn't say anything about a whale. We've just made that assumption. But there actually is a fish that could swallow a man. Confusingly, it's called a whale shark. The whale shark is not a whale, but it is a fish—the world's largest fish, in fact. And, whereas like Minnie the Moocher it may have a heart as big as a whale, unlike most sharks, it does not have big, slashing teeth, but rather plenty of small, stubby ones. Like many whales, it's a filter feeder, subsisting on plankton.

As sharks go, the whale shark is quite gentle and even playful. Divers have cavorted safely with whale sharks on many occasions. It has even been known to let divers clean its stubby little teeth. There's little doubt that Jonah could have been swallowed by a whale shark. The Bible doesn't tell us Jonah's shirt size, but it doesn't much matter. A whale shark can grow to more than forty feet long, with a mouth that would accommodate an NFL lineman who's just been through an all-you-can-eat buffet.

So, yes, when the Bible says a big fish swallowed Jonah, we have no scientific grounds to doubt it. But what happens next is a little harder to swallow. The whale shark is a great candidate for the "huge fish," except for one thing: unlike a whale, it is not an air-breather. The whale shark has five sets of gills that extract oxygen from the water. So unless Jonah had access to an aqualung, it's hard to picture how he stayed alive more than three minutes, let alone three days.

It's even harder to swallow what the Bible tells us he does next. There, in the dark, slimy, airless belly of the great fish, Jonah recites a poem.

> I cried by reason of mine affliction unto the Lord, and he heard me; out of the belly of hell cried I, and thou heardest my voice. For thou hadst cast me into the deep, in the midst of the seas; and the floods compassed me about: all thy billows and thy waves passed over me. Then I said, I am cast out of thy sight; yet I will look again toward thy holy temple. The waters compassed me about, even to the soul: the depth closed me round about, the weeds were wrapped about my head. I went down to the bottoms of the mountains; the earth with her bars was about me for ever: yet hast thou brought up my life from corruption, O Lord my God. When my soul fainted within me I remembered the Lord: and my prayer came in unto thee, into thine

holy temple. They that observe lying vanities forsake their own mercy. But I will sacrifice unto thee with the voice of thanksgiving; I will pay that that I have vowed. Salvation is of the Lord. (Jonah 2:2–9, NRSV)

Really? It's funny, you know. In writing this section, I took some pride in having remembered the whale shark and having researched its size and behavior to see if it might fit the bill for the Jonah tale. So, when I subsequently looked up the online apologetics I was disappointed to find that several apologists, including the vaunted creationist Dr. Henry Morris (founder of the Institute of Creation Research), also cite *Rhincodon typus*, as the whale shark is known.

The apologetic literature includes accounts of a dog that survived inside a whale's head, and even a man who was supposedly swallowed by a whale shark and lived for days, suffering only a permanent loss of body hair and patchy skin. After going over much the same ground, apologist Jason Tilley writes, "Absolutely I believe in the story of Jonah. I'd have to be unreasonable and illogical to doubt it given the evidence. I can't really think of a good reason to doubt it." It's funny that these anecdotes strike apologists as credible. If I were an apologist, I wouldn't want my claim of a miraculous act by God to be reenacted by some hairless, rough-skinned man. It cheapens the whole story.

Besides, can Jason Taylor and his ilk really not see its most incredible feature? Would any man, be he ever so pious, after being flung overboard and swallowed by some sort of gigantic marine monster, launch into a poetry recitation?

I mean, c'mon. Imagine that maw. I know if it were me, I'd be screaming through the last remaining air bubbles, "Get me out of here! I swear I'll catch the next donkey to Nineveh and recite poetry the whole way, but pleeeeeease God, just get me outta here!!!"

15

 Daniel's Mind-Reading Service

The biggest historical event in the Old Testament is the Babylonian conquest of Israel. Unlike the story of Pharaoh's enslavement of the Jews and their escape across the Sinai wilderness, there can be no doubt that this really happened. After a dispute over unpaid tribute, Nebuchadnezzar, the King of Babylon (now Iraq), sent his mighty army rolling into Israel. A long siege ensued, but in 587 BCE Jerusalem's walls gave way, and the Babylonian army destroyed the city, including its temple, and forced Israel's elite into exile.

This changed everything. Until then, Jews believed that their God inhabited the temple. Afterward, struggling to make sense of events, they relocated the Lord to heaven. Lost in the transition was his wife, Asherah, who was thought to live in the temple with him. Like Mrs. Lot, she got left behind, if not as a pillar of salt at least in ceramic form. Archaeologists have found numerous statues of the goddess in preexilic Israel.

Apart from losing their mother-goddess, their temple, and their land, the Israelites suffered a devastating loss of dignity. Their king was tortured and humiliated, his children were slaughtered, and the nation was torn apart. No wonder, then, that the stories which emerge from that era reflect self-doubt and an attempt to demonstrate special favor.

Case in point: the Miracle of the Vegetarian!

The Book of Daniel opens with a brief recap of the fall of Israel, noting the king and "some" of the nobility and aristocracy were brought in exile to Babylon to serve in the court of Nebuchadnezzar. According to the Jewish Encyclopedia, nearly 5,000 men were exiled, along with their wives and children. The elite among them were ensconced at the palace.

King Nebby treated his high-tone prisoners well: they got free classes in the local Chaldean language and culture, and they were served leftovers from the royal table. There was one catch, however. Like many an immigrant to America, they found themselves under pressure to change their names. You can understand why: a mere prisoner can hardly expect to hang on to a dignified handle when the king himself has to bear a ridiculous name like Nebuchadnezzar. I've written a little ditty to help you remember how to pronounce his name. It's sung to the tune of the Disney song from *Cinderella*, "Bibbidi-Bobbidi-Boo":

Nebuchadnezzar had quite a beezer,
Bibbidi-Bobbidi-Boo
Fill it with snot, and what have you got?
A sneeze that can kill with "kachoo."

So anyway, back to the Bible. We zoom in on some strapping young Israelites who are privileged to find themselves relocated to Nebuchadnezzar's HQ:

Among them were Daniel, Hananiah, Mishael, and Azariah, from the tribe of Judah. The palace master gave them other names: Daniel he called Belteshazzar, Hananiah he called Shadrach, Mishael he called Meshach, and Azariah he called Abednego. (Daniel 1:6–7, NRSV)

Before you condemn the palace master for cruel and unusual nomenclature, bear this in mind: he himself was called Ashpenaz. (Son of Razzmatazz? The Bible doesn't say.)

At any rate, Danny Boy balks at the name change and more:

Daniel resolved that he would not defile himself with the royal rations of food and wine; so he asked the palace master to allow him not to defile himself. Now God allowed Daniel to receive favor and compassion from the palace master. (Daniel 1:8–9, NRSV)

You can see how being saddled with a moniker like "Belteshazzar" could make a man bitter, but still . . . turning down royal fare for vegetables and water! Pretty drastic. At least, however, Daniel shows a commendable respect for the scientific method:

> Then Daniel asked the guard whom the palace master had appointed over Daniel, Hananiah, Mishael, and Azariah: "Please test your servants for ten days. Let us be given vegetables to eat and water to drink. You can then compare our appearance with the appearance of the young men who eat the royal rations, and deal with your servants according to what you observe." So he agreed to this proposal and tested them for ten days. (Daniel 1:11–14, NRSV)

Let's see how he and his pals fare in relation to the control group: "At the end of ten days it was observed that they appeared better and fatter than all the young men who had been eating the royal rations."

So, remember, kids: eat your spinach! As their rewards, Daniel and the gang get to continue on a strictly vegetarian diet. Oh, and God gives them "knowledge and skill in every aspect of literature and wisdom; Daniel also had insight into all visions and dreams."

This last one turns out to be a lifesaver. Nebuchadnezzar is a dreamer, but not all his dreams are good. One night the king has a humdinger of a nightmare, and he summons every enchanter in the court to find out what it portends. But Nebby is a crafty monarch. He knows those magicians are royal schmoozers, as skilled with the patter as they are with the sleight of hand and prone to tell him what he wants to hear. So, he demands that they first tell him what he dreamed and *then* interpret it. Being king, he offers them compelling incentives to obey.

> This is a public decree: if you do not tell me both the dream and its interpretation, you shall be torn limb from limb, and your houses shall be laid in ruins. But if you do tell me the dream and its interpretation, you shall receive from me gifts and rewards and great honor. (Daniel 2:5–6, NRSV)

The wisemen shuffle their feet and snivel about being asked the impossible, so the king condemns them all to a horrible death. But in the nick of time, Daniel sends a message that if the king will but stay his axe, he, Daniel, will reveal the dream and its meaning.

A little prayer, a little praise, and next thing you know Daniel is standing before the Babylonian tyrant, giving credit to God and doing the big reveal.

> You were looking, O king, and lo! there was a great statue. This statue was huge, its brilliance extraordinary; it was standing before you, and its appearance was frightening. The head of that statue was of fine gold, its chest and arms of silver, its middle and thighs of bronze, its legs of iron, its feet partly of iron and partly of clay. As you looked on, a stone was cut out, not by human hands, and it struck the statue on its feet of iron and clay and broke them in pieces. Then the iron, the clay, the bronze, the silver, and the gold, were all broken in pieces and became like the chaff of the summer threshing floors; and the wind carried them away, so that not a trace of them could be found. But the stone that struck the statue became a great mountain and filled the whole earth. (Daniel 2:31–35, NRSV)

Could anyone have pulled this off without supernatural help? It's a stretch, but the answer is a qualified maybe. Daniel just might have pulled off a monumental cold read. The cold read is a technique long used by so-called psychics and magicians to extract information from a subject while convincing him that it's a paranormal miracle.

Cold reading draws on the Barnum Effect—the term psychologists use for a tendency of people to see themselves uniquely reflected in vague generalities—provided the source is credible. Philosopher Denis Dutton writes, "experiments show quite conclusively that a Barnum description is more striking when it is believed to be derived from a credible source; they also indicate that people are often likely to be impressed by Barnum descriptions which involve some sort of arcane 'mumbo-jumbo.'"

The practitioner of the cold read begins with generalities, drawn from a knowledge of common experience, and enhanced as much as possible by hunches about the individual subject. A person's age, gender, style of dress, skin tone, hair style—all provide clues for the sophisticated cold reader. Reactions to the proffered generalities provide guidance, just as gambits in a game of chess help the grandmaster onto a sure path to victory, or "tells" help a skilled gambler know when to hold and when to fold.

The ability to cold read derives from an evolved ability in humans known as "theory of mind" or "the intentional stance." These bits of jargon refer to our ability to infer what it's like to be in someone else's sandals, to guess at their feelings and intentions. It's the supreme social skill, as it enables one to manipulate another to advantage. Of course, it varies from person to person. In some, it rises to level of high art: actors, illusionists, politicians, and con men—but I repeat myself.

That said, if Daniel's performance is a cold read, it's a first-class, grade-A, jumbo-size cold read. But consider his advantages. He knows the king, and he knows what preoccupies the king—remaining in power. Both history and the Bible tell us that Nebuchadnezzar was into erecting statues, so if Daniel knew that the king had a dream that provoked great anxiety, a crumbling statue would make a pretty good gambit. You don't need to call on Freud to see why.

But there's more: dreams are hazy, inchoate things, and our memories of dreams are unreliable. There's a good reason for this: although the science of dreams is incomplete, it's reasonably clear that dreams are byproducts of a phase of sleep in which the brain stores some memories and discards others. The narrative-production module of our consciousness—the same one that makes up false stories about what the other half of the brain is doing in patients with a severed corpus callosum—tries to weave a story out of the swirl of memories and sensory inputs that occur during sleep. Most of the time, these are unmemorable, but once in a while a dream touches a nerve, so to speak. In this case, it was a royal pain. That much Daniel has little trouble guessing. Interestingly, until you tell the story of a dream, its details are not fixed, but in this case we know that Nebuchadnezzar did all the listening. The beginning of the dream, as related by Danny Boy, is highly specific, but the ending is so vague as to truly be dreamlike. Not only might he have inferred the dream through cold-reading, but he may have convinced his mark of some details that were not there to begin with.

After all, there's no mumbo-jumbo like holy mumbo-jumbo:

"This was the dream; now we will tell the king its interpretation. You, O king, the king of kings—to whom the God of heaven has given the kingdom, the power, the might, and the glory, into whose hand he has given human beings, wherever they live, the wild animals of the field, and the birds of the air, and whom he has established as ruler over them all—you are the head of gold. After you shall arise another kingdom inferior to yours, and yet a third kingdom of bronze, which shall rule over the whole earth. And there shall be a fourth kingdom, strong as iron; just as iron crushes and smashes everything, it shall crush and shatter all these. As you saw the feet and toes partly of potter's clay and partly of iron, it shall be a divided kingdom; but some of the strength of iron shall be in it, as you saw the iron mixed with the clay. As the toes of the feet were part iron and part clay, so the kingdom shall be partly strong and partly brittle. As you saw the iron mixed with clay, so will they mix with one another in marriage, but they will not hold together, just as iron does not mix with clay. And in the days of those kings the God of heaven will set up a kingdom that shall never be destroyed, nor shall this kingdom be left to another people. It shall crush all these kingdoms and bring them to an end, and it shall stand forever; just as you saw that a stone was cut from the mountain not by hands, and that it crushed the iron, the bronze, the clay, the silver, and the gold. The great God has informed the king what shall be hereafter. The dream is certain, and its interpretation trustworthy." (Daniel 2:36–45, NRSV)

Classic fortune-telling! A little flattery, a little anxiety, and then a vague but compelling promise with no timeline. Way to go, Danno! But then the Bible goes and overeggs the pudding of credibility:

Then King Nebuchadnezzar fell on his face, worshiped Daniel, and commanded that a grain offering and incense be offered to him. The king said to Daniel, "Truly, your God is God of gods and Lord of kings and a revealer of mysteries, for you have been able to reveal this mystery!" Then the king promoted Daniel, gave him many great gifts, and made him ruler over the whole province of Babylon and chief prefect over all the wise men of Babylon. (Daniel 2:46–48, NRSV)

Really? Gifts, sure; promotion, maybe; but, falling down to worship the young Jew and his God? Uh-uh. Trouble is, Nebuchadnezzar already considered himself something of a deity. He claimed descent from the god Marduk, a longtime rival of Yahweh's. If the Babylonian king was so darn impressed with Daniel as to worship him and his God, why was Nebby's son and heir to the throne given the name Amel-Marduk, or "Marduk's man"?

It's clear that this tale is an exaggeration at best, and quite possibly a pro-Yahweh fantasy—as the very next chapter of the Book of Daniel suggests.

16

 The Fiery Furnace

Ye have learned of the miracle of the vegetarians, ye have beheld the bold cold-reader extraordinaire; now, prepare thyself for the miracle of the fiery furnace. We're still in Babylon, and Nebuchadnezzar's still the Big Kahuna (sans surfboard). Nebby turns out to be a big-time backslider as well, for he's hardly left off praising Yahweh to the heavens, but here he is erecting a golden statue of his own. And this one's no calf. The damn thing's 90 feet tall, about 80 percent of the Statue of Liberty's height.

Naturally, having gone to all the expense and trouble, Nebby wants to show off his statue. He summons everyone who's anyone to have a look. But it's more than admiration he demands:

> When they were standing before the statue that Nebuchadnezzar had set up, the herald proclaimed aloud, "You are commanded, O peoples, nations, and languages, that when you hear the sound of the horn, pipe, lyre, trigon, harp, drum, and entire musical ensemble, you are to fall down and worship the golden statue that King Nebuchadnezzar has set up. Whoever does not fall down and worship shall immediately be thrown into a furnace of blazing fire." (Daniel 3:3–6, NRSV)

Well, you know what's coming: Shadrach, Meshach, and Abednego, pious Jews that they are, refuse to bow down to the golden statue, and some tattletale astrologer rats them out. The king gives them one last chance, but they still refuse, because the God they serve offers a special extended protection plan—which, to judge from the text, is not much better than the one on offer at your local appliance store.

> Shadrach, Meshach, and Abednego answered the king, "O Nebuchadnezzar, we have no need to present a defense to you in this matter. If our God whom we serve is able to deliver us from the furnace of blazing fire and out of your hand, O king, let him deliver us. But if not, be it known to you, O king, that we will not serve your gods and we will not worship the golden statue that you have set up."
>
> Then Nebuchadnezzar was so filled with rage against Shadrach, Meshach, and Abednego that his face was distorted. He ordered the furnace heated up seven times more than was customary, and ordered some of the strongest guards in his army to bind Shadrach, Meshach, and Abednego and to throw them into the furnace of blazing fire. So the men were bound, still wearing their tunics, their trousers, their hats, and their other garments, and they were thrown into the furnace of blazing fire. Because the king's command was urgent and the furnace was so overheated, the raging flames killed the men who lifted Shadrach, Meshach, and Abednego. But the three men, Shadrach, Meshach, and Abednego, fell down, bound, into the furnace of blazing fire. (Daniel 3:16–23, NRSV)

Whew. I wonder what kind of uniforms the soldiers in Nebuchadnezzar's army had to wear? If they were made from untreated Egyptian cotton the soldiers might well have burst into flames. If they used oil-based unguents and perfumes—as many did in those days before deodorant and regular bathing—the hazard could have been even worse. But looks like someone survived the blaze:

> Then King Nebuchadnezzar was astonished and rose up quickly. He said to his counselors, "Was it not three men that we threw bound into the fire?" They answered the king, "True, O king." He replied, "But I see four men unbound, walking in the middle of the fire, and they are not hurt; and the fourth has the appearance of a god." Nebuchadnezzar then approached the door of the furnace of blazing fire and said, "Shadrach, Meshach, and Abednego, servants of the Most High God, come out! Come here!" So Shadrach, Meshach, and

Abednego came out from the fire. And the satraps, the prefects, the governors, and the king's counselors gathered together and saw that the fire had not had any power over the bodies of those men; the hair of their heads was not singed, their tunics were not harmed, and not even the smell of fire came from them. (Daniel 3:24–27, NRSV)

Once again, if you take this all literally, it is an inexplicable miracle. But should we? Could a furnace be heated to seven times its normal heat? How? And how would anyone know? They didn't have thermometers in those days. The detail about the furnace being so hot that it killed the soldiers who shoved the blessed trio into the furnace is just a bit much. Be he ever so loyal, no soldier is likely to stand in the face of deadly heat long enough to be killed by it.

Since radiative heat dissipates with the inverse square law (dropping to one-fourth the intensity at twice the distance), there's no way that what killed those outside would spare those inside the furnace, or that clothing would be neither scorched nor imbued with smoke.

But these details look very much like storyteller's gilding applied to make the story sparkle. Indeed, the whole story sounds much like a myth, but assuming it sprang from some incident, what might have happened? Could anyone enter a furnace and live?

Well, perhaps. Furnaces in the ancient Middle East were not up to the standards of Pittsburgh Steel. They relied on charcoal in a campfire-style bed. The charcoal would be heated by pipes inserted into the fire, which could be blown through or attached to bellows. But even with the most vigorous bellowing, they never hit 2,800 degrees Fahrenheit, the melting point of iron. Instead, the ancients relied on a process called iron bloom, in which the carbon coming off the charcoal would soften the iron so that it could be worked. As you'll see below, even with bellows, Nebuchadnezzar's furnace would have topped out at about 1,800 degrees Fahrenheit. If that was seven times hotter than usual, then on most days the mighty Nebuchadnezzar couldn't even bake a pizza in his furnace.

What's more, it's likely to have been outdoors. There were ancient furnaces with walls, but they weren't large enough to accommodate people. Nobody built room-sized blast furnaces in those days. Ancient versions were more like the backyard chiminea that are so popular in the

Southwest. Babylonian furnaces had to be small, because users needed to grasp lumps of iron with tongs and pull them out when heated. So, if anything, the furnace in question was likely an outdoor fire pit lined with wood or charcoal.

That raises the interesting possibility that Shadrach, Meshach, and Abednego may have been firewalkers. After all, the Bible tells us that they were hanging around court magicians—maybe they picked up the trick from one of their Persian or Indian companions. It also quotes Nebby as saying he saw the men "walking around in the fire, unbound and unharmed."

Firewalking on charcoal is an ancient practice thought to have originated in India. The oldest reference to it there comes in a story dated to about 1,200 BCE. From there it spread, er, like wildfire. To understand how something as idiotic as walking across live coals could have gotten started, you only have to watch a YouTube video of an adolescent male putting a lit firework up his backside. Guys do dumb stuff.

Truth is, though, firewalkers take far less risk than teenage butt-stuffers. Firewalking became something of a paranormal fad in the United States during the 1980s, when self-styled gurus and, even worse, corporate retreat leaders, earned big bucks getting people into the right frame of mind to survive a walk across the coals.

Intrigued by the claims and counterclaims, physicist David Willey of the University of Pittsburgh decided to find out for himself. He observed the world's hottest firewalk, in Redmond, Washington, on October 18, 1997. More than 20 people walked over a bed of coals whose temperature, boosted by a leaf blower, ranged from 1,602 degrees to 1,813 degrees Fahrenheit. No one was harmed. Many walked the 11.5-foot course more than once. The following summer Dr. Wiley took part in the longest (165 feet) firewalk and experienced for himself the remarkable but nonmiraculous paradox of being able to walk safely across a hot fire.

He notes that heat can be transmitted in several ways, starting with radiation. Our faces are especially sensitive to radiative heat, such as strong sunlight, so standing a few feet from the fire can be painful. However, once there is a coat of ash on the surface of a coal, it blocks the radiative heat from the sole. What's more it also insulates the foot from that other relevant form of heat transmission, conduction.

Here's an excerpt from his account:

> Conduction happens when energetic molecules, the hot coals, that are vibrating collide with more sedate molecules, the soles of the feet, thereby transferring energy to them, but the thermal conductivity of coarse charcoal is very small and that of skin or flesh is only about four times more. By comparison the thermal conductivity of most metals is several thousand times larger . . .
>
> It would seem then, that a firewalk of short length is something any physically fit person could do and that it does not need a particular state of mind. Rather, it is the short time of contact and the low thermal capacity and conductivity of the coals that is important, and it is not necessary for the feet to be moist nor callused, although either may be of slight benefit. Longer walks appear to be possible if a layer of insulating ash is allowed to build up on a well packed down bed, where the temperature has been allowed to fall significantly from what it was when the coals were at their hottest.

Dr. Wiley, having safely traversed the 165-foot-long firebed in his bare feet, concludes, "firewalking is understandable in terms of basic physics and is not supernatural nor paranormal."

Now, to be sure, this story of Shadrach, Meshach, and Abednego does not perfectly align. There's a mysterious stranger to account for. But perhaps the strangest thing about the tale is not their survival in an ancient furnace or the flickering figure who joins them but this conundrum: if they were so pious, why, unlike Daniel, did they give up their Jewish names and accept those ridiculous Chaldean monikers?

I mean, Abednego? Sounds practically racist.

17

 Daniel in the Lion's Den

Despite his pretensions to living god status, Nebuchadnezzar proved to be mortal after all—as did the kings who followed him. Several reigns later, though, something truly amazing happens. The Babylonians, conquerors of Israel, are themselves conquered. In telling the tale, the Bible gives us one of the most durable clichés in history: "the handwriting is on the wall."

King Belshazzar is holding a feast when "the fingers of a human hand appeared and began writing on the plaster of the wall of the royal palace. . . . Then the king's face turned pale, and his thoughts terrified him. His limbs gave way, and his knees knocked together." If it seems a bit comical, no wonder. Scholars have identified this passage as satirical. But things turn grim when Daniel, now a graybeard courtier, comes in and interprets the writing for the king: the royal bucket is about to be kicked. Indeed, a Persian-backed alliance arrives at the palace gates, and before dawn Belshazzar is done.

Darius the Mede takes the throne, and courtiers rush in to flatter the new king. They persuade him to sign an anti-piety decree that "whoever prays to anyone, divine or human, for thirty days, except to you, O king, shall be thrown into a den of lions."

Of course, Daniel, pious Jew that he is, finds himself entrapped by his enemies and hauled in on a charge of illegal praying three times a day!

Now, here's where things take an odd turn. You might think that a king who'd just signed a decree directing all prayers to him would be all for it. But no. Darius gets all wobbly and wants to reprieve Daniel.

"When the king heard the charge, he was very much distressed. He was determined to save Daniel, and until the sun went down he made every effort to rescue him."

Hey, no problem, your majesty. Just repeal or amend the decree. But not so fast. "Then the conspirators came to the king and said to him, 'Know, O king, that it is a law of the Medes and Persians that no interdict or ordinance that the king establishes can be changed.'" Wow. Good thing they didn't pass Prohibition.

Well, somehow dopey Darius swallows this ludicrous restriction on his kingly power and personally seals Daniel in the lion's den. Even though the king's own edict states that everyone must worship him and only him, he offers Daniel this generous blessing, "May your God, whom you faithfully serve, deliver you!" Then, the Bible tells us, he went to his room and spent a sleepless night fasting. Good thing he didn't have a Twitter account.

Well, if all that were not miracle enough, when day breaks the den is opened and there is Daniel unharmed and full of good wishes for the man who just made him spend the night with ravenous beasts.

King Darius returns the favor with interest. Though just a day before, he was punctiliously bound by the law, today he simply issues an arbitrary command, sweeping the guilty and innocent alike to their doom. "The king gave a command, and those who had accused Daniel were brought and thrown into the den of lions—they, their children, and their wives. Before they reached the bottom of the den the lions overpowered them and broke all their bones in pieces."

That will teach those wives and children! But wait, there's more:

Then King Darius wrote to all peoples and nations of every language throughout the whole world: "May you have abundant prosperity! I make a decree, that in all my royal dominion people should tremble and fear before the God of Daniel: For he is the living God, enduring forever. His kingdom shall never be destroyed, and his dominion has no end. He delivers and rescues, he works signs and wonders in heaven and on earth; for he has saved Daniel from the power of the lions." So this Daniel prospered during the reign of Darius and the reign of Cyrus the Persian. (Daniel 6:25–27, NRSV)

Could a man survive a night in a den of lions? Kings who kept lions generally went in for male lions, whose manes are a symbol of kingly puissance. But male lions, we now know, are lazy bastards who sit around waiting for others to do the hard work and bring the fruits of their labor to them. Much like real kings. Still, male lions can be murderous. Again, like real kings. So, surviving in a den of them takes some good luck and maybe a few chicken breasts concealed in your robes to toss to the lions as placatory gifts.

Daniel and the lion's den makes a great story. However, the "miracle" of surviving does not require supernatural intervention. It's the rest of the story that defies belief. Darius and Cyrus were real people, and there's no hint that they ever bowed down to Yahweh. They did, however, allow the Jews to return to their homeland, where pious legends like this could be safely told, retold, and embellished.

18

 The Virgin Birth

One of the central miracles of the Bible, especially for Roman Catholics, is the immaculate conception of Jesus. For such an important event, reports in the Bible are rather spotty. Only two of the four gospels mention it, and Paul, who has an awful lot to say about Jesus, overlooks it entirely. So does the remainder of the New Testament. All the same, it is a key miracle of the Bible, so let's plunge in.

In Luke, the event comes equipped with its own prophecy:

> [T]he angel Gabriel was sent by God to a town in Galilee called Nazareth, to a virgin engaged to a man whose name was Joseph, of the house of David. The virgin's name was Mary. And he came to her and said, "Greetings, favored one! The Lord is with you." But she was much perplexed by his words and pondered what sort of greeting this might be. The angel said to her, "Do not be afraid, Mary, for you have found favor with God. And now, you will conceive in your womb and bear a son, and you will name him Jesus. He will be great, and will be called the Son of the Most High, and the Lord God will give to him the throne of his ancestor David. He will reign over the house of Jacob forever, and of his kingdom there will be no end." Mary said to the angel, "How can this be, since I am a virgin?" The angel said to her, "The Holy Spirit will come upon you, and the power of the Most High will overshadow you; therefore the child to be born will be holy; he will be called Son of God. (Luke 1:26–35, NRSV)

And yet, apparently on God's instructions, the parents gave him the everyday name of Joshua (or Yeshua, later Greco-Latinized as Jesus). It's a bit odd, since there's already been a major figure in Jewish history named Joshua (the genocidal maniac who succeeded Moses). But there's a far more important figure with whom Jesus will be associated: King David. Yet, rather than name their son Davy, the parents call him Josh. Go figure.

We'll get to the bigger mystery here (what does the Bible mean by a "virgin birth"?) momentarily, but first this name thing needs a little more digging. The Gospel of Matthew tells us that an angel appears to Joseph in a dream to explain why his fiancée is pregnant, and right away the connection to Big D is made: "Joseph, son of David, do not be afraid to take Mary as your wife, for the child conceived in her is from the Holy Spirit. She will bear a son, and you are to name him Jesus, for he will save his people from their sins."

A little etymology may help. The name "Joshua," according to Dr. James Strong's 1890 classic research, means "Jehovah is salvation." Well, that clears it up. Only not quite. "Jehovah" is an anglicized way of saying "Yahweh," but Yahweh is not Jesus' dad—at least, according to some scholars. This isn't the place—and I'm not the guy—to sort out all the issues and controversies, but to avoid a paternity suit let me just sketch a few reasonably well-accepted propositions.

Yahweh is a particular god, the one we met a few chapters ago leading the Jews through the Sinai and on to slaughter the Hittites, Perizzites, Fezziwigs, and whatnot. But before Yahweh came El. This is a bit awkward for monotheists, but it seems there were numerous gods running around the Holy Land. It's only after the exile to Babylon that monotheism begins to prevail in Judaism. The Bible mentions dozens of gods, including such favorites as Baal, Marduk, Beelzebub, as well as a mysterious "Queen of Heaven," who may be Asherah, aka Mrs. God (see chapter 15).

But towering over them all was El, the proto-deity, head of the Canaanite pantheon, and also a generic god to various cultures, including at least some of the Jews. Remember, there were twelve tribes of Israel, but, Mosiac myth aside, no central authority like a pope to sort out the deity puzzle. El, also known as Ila (later rendered "Allah" in Arabic), not only appears in the Bible, but is also clearly a father. In the authoritative Hebrew version, known as the Masoretic text, El appears more than 200 times. But so does "Elohim," which in Hebrew is a masculine plural, indicating that there are El juniors running around.

All this and more lead some students of the Bible to see a higher hierarchy here. "All the texts in the Hebrew Bible distinguish clearly between the divine sons of Elohim/Elyon and those human beings who are called sons of Yahweh," writes Margaret Barker in her book, *The Great Angel: A Study of Israel's Second God*. This leads Neil Godfrey to assert: "Yahweh was one of the sons of El Elyon; and Jesus was also in the Gospels described as a Son of El Elyon, God Most High."

But none of this can be considered conclusive, since in many other parts of the Bible, "El" and "Yahweh" are used interchangeably, and scholars disagree about whether they were ever considered separate gods. Still, it cannot be denied that the Bible holds intriguing passages like this, opening the Psalm of David: "Ascribe to the Lord, O heavenly beings . . . " In the original Hebrew, those "heavenly beings" are unambiguously the sons of El.

But, let's put all that to one side and agree that God the Father fathered Jesus. What exactly is that supposed to mean? Did the Holy Spirit convey God's haploid gametes—that is, sperm—into Mary's womb?

Inconceivable! For one thing, the Ancient One is far too old for her. For another, in keeping with his status as the Supreme Being, he would have to have *perfect* DNA. But what could that mean? The whole point of evolution is that the stock of genes evolves. It adapts over time to a changing environment. At one moment in natural history it pays to be a dinosaur, at another, a shrew. What would be in the DNA of an infinite being? The perfect allele for every occasion? ("Darling, I feel like fur tonight.")

But in that case, how would God's titanic sperm fuse with Mary's oocyte? Information is physical, and efficient an encoder as DNA may be, each gene takes up space. It's a conundrum. Just who is that proud poppa?

Perhaps delivering divine sperm is not only impractical but beneath the dignity of God. It is, after all, shocking to contemplate the Almighty producing a sample. So, maybe we are better off presuming that the Holy Spirit simply flew in with a specially prepared set of 23 "Christ chromosomes."

Even then, there are challenges. Mammalian conception is a fantastically complicated molecular dance. To fertilize Mary's egg, the Holy Spirit would have had to bring along several molecular keys. One of them, a protein known as DE, is normally expressed on the surface of a sperm. Without it, docking doesn't take place. Another set of proteins with the wonderful acronym ADAM also seems to be necessary for proper fusion. The whole process needs exquisite timing and sequencing, and more often than not it fails. (That's why it usually takes time to get pregnant.)

If God is the biological father of Jesus, then perhaps Mary was impregnated by what we now call artificial insemination. Today, it's possible to bypass the natural steps and employ an intracytoplasmic injection technique to bring the two sets of DNA together. This is normally done outside the womb, and then the egg is incubated for at least 12 hours before implantation. So, one way or another the Holy Ob-Gyn would have had to make one heck of a house call to bring it off.

That's impressive, but is it a miracle? Only in the time and place where it occurred. Since the birth of Louise Brown in 1978, *in vitro* fertilization (IVF) has become commonplace—though Catholic doctrine condemns it.

The idea that Jesus was a test-tube baby troubles some Catholics. In an online forum called Catholic Answers, a junior member writes, "If God did it, why is it immoral for us to do? . . . God separated the Marital act from the Procreative act . . . which is a clear violation of the Catechism."

It will doubtless come as huge relief to such troubled souls to learn that there's an out for God, a way in which the Everlasting could have dodged the sin of extramarital insemination. For a truly "immaculate" conception, the Holy Spirit could have sparked parthenogenesis in Mary.

Parthenogenesis is defined as reproduction without fertilization. It happens when an egg gets tired of sitting around waiting for a sperm and doubles its own genetic material. The egg can then develop into an embryo and occasionally result in a live birth. This phenomenon has been observed in a wide range of creatures from aphids to sharks to the occasional turkey, but it is quite rare in mammals.

Since parthenogenesis puts 100 percent of a mother's genes into the next generation, you might wonder why females bother with sex at all—especially if you've been married for a while. Biologists believe that the evolutionary function of sex is to mix up genes so as to weed out deleterious mutations and stay ahead of parasites.

Parthenogenesis would seem to be a good candidate for explaining Mary's pregnancy. It even means, in the original Greek, "virgin birth." In human development, however, sperm are more than a UPS delivery truck for DNA. The sperm also helps orchestrate the sequence of postfertilization cell specialization. In cases where an egg has been artificially prompted to start dividing, a crazy jumble of cells results, with bits of different organs all mashed together. The sad little lump dies within days.

No human birth has ever been known to result from parthenogenesis. But that does not mean that it cannot. Indeed, though odds are hugely against, it may have happened without anyone being the wiser. Only genetic testing could confirm such a thing.

But there is one more stumbling block. So far as we know, if that's how Mary became pregnant, she would necessarily have given birth to a baby girl. Only a sperm can furnish a Y chromosome. Which raises the question . . . could Jesus have been transgender? The pope would plotz!

Before leaving this topic, we should note that some controversy exists over whether "virgin" was meant at all. The Hebrew word, some argue, has an ambiguity that could depict Mary simply as a young woman. However, the narratives as we find them seem pretty clear on this point, and its importance remains easy to grasp. In a time when DNA tests were unavailable, virginity was a husband's only guarantee of paternity, so virginity mattered to men. Hence, Joseph's consternation:

> Now the birth of Jesus the Messiah took place in this way. When his mother Mary had been engaged to Joseph, but before they lived together, she was found to be with child from the Holy Spirit. Her husband Joseph, being a righteous man and unwilling to expose her

to public disgrace, planned to dismiss her quietly. (Matthew 1:18–19, NRSV)

On the other hand, if the virgin birth of Jesus were so all-fired important, we come full circle to this question: why is it only mentioned in Matthew and Luke? And maybe in John 8:42:

> Jesus said to them, "If God were your Father, you would love me, for I came from God and now I am here. I did not come on my own, but he sent me. Why do you not understand what I say? It is because you cannot accept my word."

Apologist Don Stewart sees this passage as evidence of Jesus' divine birth. But surely it is at least as reasonable to read it as evidence that Jesus had a stain on his family history that needed scrubbing in the Gospels.

Géza Vermes, a professor of Jewish Studies at Oxford, notes, "hostile Jewish and pagan gossip rumoured that Jesus was conceived out of wedlock, and that his father was Panthera, a Roman soldier."

It is important to bear chronology in mind. The consensus among biblical scholars holds that the earliest of the gospels, Mark, was written no less than forty years after the death of Jesus. (Among the evidence: Mark refers to the Roman destruction of the Temple, which happened in 70 CE.) The letters of Paul, which come closest to the life of Jesus, were written some twenty years after the crucifixion.

We can be certain that for decades after his death, only small bands of people accepted Jesus as the Messiah. It is therefore entirely possible that the aspersions about his paternity came long after the death of Jesus.

If so, his defenders would have searched for a way to turn the insult around. They may have found it in Isaiah, where it is written:

> "Hear then, O house of David! Is it too little for you to weary mortals, that you weary my God also? Therefore the Lord himself will give you a sign. Look, the young woman [or "virgin" in Greek] is with child and shall bear a son, and shall name him Immanuel ["God is with us"]. (Isaiah 7:13–14, NRSV)

Is there any other reason to think that the tale of the virgin birth was a defensive move? Professor Vermes thinks so. Noting that the narratives of Jesus' career make no reference to his origins, he infers: "The infancy narratives are best understood as late additions to Matthew and Luke."

So, it may be that the authors of those two gospels saw the chance to turn the sow's ear of bastardy into the silk purse of divine paternity.

19

 The Star of Bethlehem

When I was a child, as the holiday season rolled around we sang Christmas carols in school. Even then, I remember thinking it was a little weird for the kids in class who weren't Christian. But then I was living in Egypt, a country that's 90 percent Muslim, and the refuge to which, according to the Gospel of Matthew, Joseph and Mary fled with the infant Jesus when Herod launched a baby-killing spree. No, I'm not Egyptian, and I'm certainly not a secret Muslim. I was born in California to parents whose families were Christian as far back as anyone knows. But when I was four my dad got a job teaching at the American University in Cairo, so there we were.

At any rate, at the international school I attended, there was one song that even the most culturally sensitive boys at the back of the room could sing with no qualms. It went something like this:

> *We three kings of Orient are*
> *Puffing on a rubber cigar.*
> *It was loaded and exploded.*
> *That's how we traveled so far . . .*

We didn't know why anyone would smoke a rubber cigar. If there was any sexual innuendo in the lyrics it sailed safely over our heads. But the mental image of an exploding cigar? That cracked us up.

All too soon our subversive moment passed, and we had to join the good kids in singing the refrain:

Oh . . . oh Star of wonder, star of night
Star with royal beauty bright
Westward leading, still proceeding
Guide us to thy Perfect Light

So we were left to wonder along with the rest of the curious . . . what was this "star"? The carol is not much help. Written in 1857 by Rev. John Henry Hopkins, it can hardly be considered an authoritative guide to history.

For the real story, we turn now to Matthew 2:

> In the time of King Herod, after Jesus was born in Bethlehem of Judea, wise men from the East came to Jerusalem, asking, "Where is the child who has been born king of the Jews? For we observed his star at its rising, and have come to pay him homage." When King Herod heard this, he was frightened, and all Jerusalem with him; and calling together all the chief priests and scribes of the people, he inquired of them where the Messiah was to be born. They told him, "In Bethlehem of Judea; for so it has been written by the prophet: 'And you, Bethlehem, in the land of Judah, are by no means least among the rulers of Judah; for from you shall come a ruler who is to shepherd my people Israel.'" (Matthew 2:1–6, NRSV)

More on the star in a moment, but let's pause to note several interesting features of this narrative. First, no way are these guys "kings." They are, to use the Greek term, *magi* from the East, which at that time would have been Babylon, Persia, or India. Since they can evidently speak the local lingo, Babylon is our best bet.

To be wise, in those days, likely meant being an astrologer. Astronomy and astrology—the Western versions at least—were born as conjoined twins in Babylon. Astronomy is a science, astrology an elaborate superstition. In its earliest form, astronomy consisted of observing the apparent clusters of stars and their apparent shifts over the course of a year, and then relating these to agricultural cycles—seasons, floods, dry spells, etc.

Astrology is an attempt to relate the same patterns to the destinies of individuals. This belief system is problematic in several ways. First, completely independent systems of astrology exist—the Chinese and Babylonian ones have no meaningful connections. A characteristic of real science is that it is consistent across cultures. The speed at which an apple (persimmon, coconut, kiwi, or what-have-you) falls toward the center of the Earth when dropped from a tree is virtually the same all over the planet (though it can be measured in culturally determined units). The fate that astrologers read out of patterns in the night skies, on the other hand, has no such consistency. Were you born under Leo the Lion? Or the Rat? Depends on who you ask.

Astrology presumes you are integrated with the whole universe in a way intimately related to your birthdate, and that astrologers can interpret that system for you. But we now know that there are vastly more stars than the 2,000 or so visible to the naked eye. Our Milky Way galaxy alone contains at least 100 billion stars, and it's just one of trillions of galaxies. What input do they have? Who's keeping track of that? Not NASA.

Third and most damning, the stars that allegedly influence human affairs are at widely varying distances. The familiar Big Dipper, for example, includes one star that is just 63 light-years away and another that is 210 light-years away. Information cannot travel faster than the speed of light. This is a law of nature that Einstein discovered and many have subsequently confirmed. It means that there is no way for the stars of the Big Dipper or most any other constellation to coordinate their influence on the scale of a human lifetime.

With scientific information now readily available, astrology is for suckers. But back in the day, everyone who was anyone made use of astrologers. No king would leave home without one. So, if you were claiming to be a wise man from Babylon, that would be the field for you.

And, indeed, the Magi sound like astrologers: just listen to what they say: "we observed his star at its rising." If they were wise about, say, life insurance, they'd hardly be wasting their time trying to assign heavenly bodies to promising newborns. Instead, they'd be talking affordable term life policies, contingent beneficiaries, and stuff like that.

But as it happens, these three astrologers from Babylon aren't all that wise about risk assessment. They stop at the royal court in Jerusalem to ask directions. And here's the really interesting thing: they are privy to a prophecy that Yahweh has given the Israelites—it's a prophecy about the coming of the Christ! Just the thing to blab about to King Herod.

In a plot device that every scriptwriter will recognize as "explaining to the audience," Herod plays dumb. "So," he says to his advisers, "this here Messiah who's gonna overthrow me and rule Israel . . . where's he supposed to be born again?"

All Together: "Bethlehem!" (1930s Screwball Comedy Translation)

Back to the story:

> Then Herod secretly called for the wise men and learned from them the exact time when the star had appeared. Then he sent them to Bethlehem, saying, "Go and search diligently for the child; and when you have found him, bring me word so that I may also go and pay him homage."

Yeah, right. Homage.

> When they had heard the king, they set out, and there, ahead of them, went the star that they had seen at its rising, until it stopped over the place where the child was. When they saw that the star had stopped, they were overwhelmed with joy. On entering the house, they saw the child with Mary his mother; and they knelt down and paid him homage. Then, opening their treasure chests, they offered him gifts of gold, frankincense, and myrrh. And having been warned in a dream not to return to Herod, they left for their own country by another road. (Matthew 2:7–12, NRSV)

What could this star that went ahead of them and stopped over the manger have been? One thing is for sure: we can't read this literally. A new star in the vicinity of the Holy Family would have incinerated them—not to mention destroying our solar system. So, we're left with two possibilities: either a luminous object in their line of sight that indicated the direction and then disappeared ("stopped"), or . . . it's a miracle!

The apologist web site ChristianAnswers.net opts for the latter:

The star went before the Magi and led them directly from Jerusalem to Bethlehem. This is a distance of about six miles, in a direction from north to south. However, every natural object in the sky moves from east to west due to the Earth's rotation. It also is difficult to imagine how a natural light could lead the way to a particular house. The conclusion is that the Star of Bethlehem cannot be naturally explained by science! It was a temporary and supernatural light.

The first point is a good one: Bethlehem is indeed south of Jerusalem. But the conclusion does not necessarily follow. There could be other explanations. Someone might have constructed a sky lantern, for example. A sky lantern is a paper hot-air balloon powered by a candle. It was invented in China hundreds of years before the birth of Jesus. The idea could have made its way to Jerusalem and drifted to Bethlehem. Or, a glowing alien spacecraft could have led the wise men on their way. I'm not saying that's what happened; I'm only pointing out that stating "not x, therefore supernatural" is never a valid argument.

But is it really not x? The claim that all natural objects in the sky move east to west is inaccurate. For observers in the northern hemisphere, for instance, Polaris, the North Star, barely moves at all. Even more conveniently, for those in the Middle East, Polaris hugs the horizon. However, it is in the wrong direction. As it happens, there is no visible "South Star."

Could there have been a temporary one in place for the "wise men"? You betcha. There are at least two good candidates. One would be a comet. Comets are big, dirty snowballs left over from the formation of the solar system. There are about 4,000 known comets, but the Oort Cloud, out beyond Pluto, may contain billions. Some are in a stable orbit, but others get jostled and make a one-time trip around the Sun— from any direction. As they approach and heat up, they become visible for a while. Then, conveniently, they stop being visible.

Anything else do that? A supernova. Aging Red Giant stars may be completely invisible to the naked eye. But when one exhausts its fuel, suddenly the weight of all its mass causes it to collapse. Even if it's stellar, you can only pack so much into a suitcase; much of the star stuff bursts out in a colossal explosion. The flash is so bright that occasionally a supernova is visible in daylight. It can remain visible for months. And, by cracky, it just so happens that Chinese astronomers recorded at least two supernovae around the time of Jesus' birth. At least one of them might have served as a "This way to Bethlehem" pointer.

There is yet another possibility, one that may not satisfy the "lead from north to south" idea, but which does have some other advantages. Bear in mind that if the Magi were prompted to hit the road for Jerusalem by the appearance of a star, they had a journey of more than 500 miles ahead of them. That would have been roughly a month-long trek in those days.

Since 1976 Professor David Hughes, an astronomer from the University of Sheffield, has been arguing that what the Magi saw was not a star, nor a comet, but a triple conjunction of Jupiter and Saturn. Within a short period of time, the two largest planets crossed within the line of sight of observers on Earth. "This happens when you get an alignment between the Sun, the Earth, Jupiter, and Saturn," Hughes told the BBC.

The argument turns on timing. "If you read the Bible carefully," Hughes said in his BBC interview, "the Magi saw something when they were in their own country, so they traveled to Jerusalem and had a word with King Herod." Then, he noted, "when they left Jerusalem [for] Bethlehem, they saw something again."

As the great science and math writer Martin Gardner pointed out, Hughes is recapitulating an argument first set forth by the even greater Johannes Kepler, a seventeenth-century astronomer who was also a devout Christian. But of course in his day Fundamentalism had yet to be invented.

Could the "Star of Bethlehem" have been an episode in the celestial *pas de deux* of our neighboring gas giants? Possibly, though you'll never convince the good folks at ChristianAnswers.net. Personally, I find the supernova explanation more plausible. But all explanations pale, in my opinion, next to the argument that this is just an oral legend that the author of Matthew pinned to the tale of Jesus' birth to give it more pizzazz. Why?

First, it's doubtful that Jesus of Nazareth was born in Bethlehem.

This would be like claiming that John Adams, that stout New Englander of Quincy, Massachusetts, was brought into the world in Charleston, South Carolina. You'll recall that Bethlehem lies south of Jerusalem; well, Nazareth is a good 80 miles to the north, but culturally even more distant than that. So, what would his parents be doing so far from home at a time when Mary was due to give birth?

The Gospel of John provides a clue. Philip, one of the first to be convinced that there's something special about Jesus, is trying to persuade Nathanael to join the entourage. He says, "We have found him about whom Moses in the law and also the prophets wrote, Jesus son of Joseph from Nazareth." Unimpressed, Nathanael replies, "Can anything good come out of Nazareth?" (John 1:45–46, NRSV)

Apparently, the hometown of Joseph, Mary, and Jesus had all the glamour of Newark. Even a manger in Bethlehem was classier than a hospital room in Nazareth. Besides, Bethlehem was the place to be if you wanted your kid to fulfill prophecy. As Bishop John Shelby Spong and many other scholars have pointed out, the portrayal of Jesus fulfilling prophecies does not occur evenly across the Gospels. A trend is discernible: the later the gospel the more elaborate the back-filling of prophecy. The Bethlehem story may well represent another attempt to fit the narrative of Jesus' life to preexisting prophetic scriptures.

As Gardner notes, a star marking the birth of a Jesus is a well-worn trope, "similar to many ancient legends about the miraculous appearance of a star to herald a great event, such as the birth of Caesar, Pythagoras, Krishna (the Hindu savior), and other famous persons and deities. Aeneas is said to have been guided by a star as he traveled westward from Troy to the spot where he founded Rome."

But there's another, more troubling reason that makes me doubt the story of the Star of Bethlehem. It's those wise men. Remember, they are from Babylon. Why should they care about the future ruler of a backwater like Israel? This would be like three male Hollywood celebrities—say, Sean Connery, Paul Newman, and Steve McQueen—dropping everything on December 25, 1971, to walk to Canada with presents in their backpacks for baby Justin Trudeau. Yes, Prime Minister Justin Trudeau was born on Christmas Day (and Jesus almost certainly wasn't).

Okay, Canada is farther (and colder), and the wise men didn't have that kind of star power. Maybe I'm exaggerating. But only a little.

What was in it for the Magi? Nothing but sand fleas and blisters.

20

 Introducing . . . Jesus the Exorcist

Once the birth of Jesus is accounted for, his biography goes blank for a long time. Sure, there's the flight to Egypt, but the Savior is still in Pampers at that point. And there's a puzzling contradictory note on his infancy in Luke: there, we are told, that instead of hot-footing it to Egypt to await Herod's death, Joseph took his wife and 40-day-old baby right into the hot zone, Jerusalem itself, for a ritual at the Temple:

> When the time came for their purification according to the law of Moses, they brought him up to Jerusalem to present him to the Lord (as it is written in the law of the Lord, "Every firstborn male shall be designated as holy to the Lord"), and they offered a sacrifice according to what is stated in the law of the Lord, "a pair of turtledoves or two young pigeons." (Luke 2:22–24, NRSV)

Remember, now, that according to Matthew, King Herod the Great was so afraid of being upstaged by a usurping Messiah that he had all the firstborns in the land slaughtered. But here are Joseph and Mary presenting their *firstborn son* for a blessing at the very temple that Herod is remodeling. If that weren't enough, a blabbermouth old geezer named Simeon draws attention to the child in a way calculated to send Herod's paranoia right up through the temple mount.

> Now there was a man in Jerusalem whose name was Simeon; this man was righteous and devout, looking forward to the consolation of Israel,

and the Holy Spirit rested on him. It had been revealed to him by the Holy Spirit that he would not see death before he had seen the Lord's Messiah. Guided by the Spirit, Simeon came into the temple; and when the parents brought in the child Jesus, to do for him what was customary under the law, Simeon took him in his arms and praised God, saying, "Master, now you are dismissing your servant in peace, according to your word; for my eyes have seen your salvation, which you have prepared in the presence of all peoples, a light for revelation to the Gentiles and for glory to your people Israel." And the child's father and mother were amazed at what was being said about him. Then Simeon blessed them and said to his mother Mary, "This child is destined for the falling and the rising of many in Israel, and to be a sign that will be opposed so that the inner thoughts of many will be revealed—and a sword will pierce your own soul too." (Luke 2:25–35, NRSV)

You know . . . the light touch. He's backed up by an old crone called Anna (for once, a minor female character in the Bible has a name!). Miraculously, none of Herod's spies reports this, and the Holy Family (which is simultaneously taking refuge in Egypt), departs in peace. The story then goes blank for a dozen years.

When they had finished everything required by the law of the Lord, they returned to Galilee, to their own town of Nazareth. The child grew and became strong, filled with wisdom; and the favor of God was upon him. (Luke 2:39–40, NRSV)

Being told "you're so special" all the time isn't necessarily the best thing for a child. At any rate, by the time he hits puberty it seems to have fostered a touch of arrogance in Jesus. To be fair, though, his parents are annoyingly dense. From the moment of his conception, they've been told over and over that he's the Messiah, the fulfillment of prophecy, the freakin' son of God, but somehow it keeps slipping their minds:

Now every year his parents went to Jerusalem for the festival of the Passover. And when he was twelve years old, they went up as usual for the festival. When the festival was ended and they started to return, the boy Jesus stayed behind in Jerusalem, but his parents did not know it. Assuming that he was in the group of travelers, they went a day's journey. Then they started to look for him among their relatives and friends. When they did not find him, they returned to Jerusalem to search for him. After three days they found him in the temple, sitting among the teachers, listening to them and asking them questions. And all who heard him were amazed at his understanding and his answers. When his parents saw him *they were astonished*; and his mother said to him, "Child, why have you treated us like this? Look, your father and I have been searching for you in great anxiety." He said to them, "Why were you searching for me? Did you not know that I must be in my Father's house?" *But they did not understand what he said to them.* [Emphasis added.] (Luke 2:41–50, NRSV)

Geez, mom and dad, how many times do we have to tell you? Get a grip!

Anyway, that's all the data we have on young Jesus. Did he go to school? We don't know. Scholars hotly debate whether Jesus was literate or not, but there's scant evidence for either position. Did he have a Bar Mitzvah? If so, did he get a bunch of crummy neckties, or maybe some *denarii*, the coin of the realm? We don't know.

We leap over that question to examine the career of Jesus in his prime. This is tricky, because as we've already seen, biblical accounts differ, sometimes radically, from gospel to gospel. For starters, let's turn to the earliest of the Gospels, the Book of Mark. Scholars reckon this tale of the Messiah was set down about 70 CE, or some 40 years after the death of Jesus. It could not have been written earlier, they say, because it shows awareness of the Roman destruction of the Jewish temple, which took place that year.

Another reason for placing this gospel first: Mark is a comparatively slim volume, with relatively modest miracles in it. Still, the eponymous author wastes little time in getting to them. Just one preliminary: Jesus needs an entourage.

According to Mark, the Messiah swings by the Sea of Galilee to round up some disciples. These are no jobless loafers but men gainfully employed in the fishing trade.

As Jesus passed along the Sea of Galilee, he saw Simon and his brother Andrew casting a net into the sea—for they were fishermen. And Jesus said to them, "Follow me and I will make you fish for people." And immediately they left their nets and followed him. As he went a little farther, he saw James son of Zebedee and his brother John, who were in their boat mending the nets. Immediately he called them; and they left their father Zebedee in the boat with the hired men, and followed him. (Mark 1:16–20, NRSV)

Now, this is strange. A believer, reading with foreknowledge that Jesus is the Christ, might think, "Well, sure, who wouldn't leap at the chance to work for the Messiah?" But as the next section makes clear, Jesus hasn't, er, come out yet. So, think of it from the point of view of these dudes. There they are making a decent living with their nets when along comes a stranger who tells them to give up their livelihoods to follow him. Why? To become "fishers of men." Would you go? On the spot? Yet, so attractive is the offer that James and John leave their poor father, Zebedee, adrift in his boat with the hirelings.

All to become "fishers of men." What could that mean? Presumably, to help save souls. But as we'll see, the apostles may have thought they were signing on for something more akin to a traveling medicine show.

The other, later gospels have slightly different orders of events. Luke has Jesus preaching in Galilee before he seeks disciples, so that his reputation might have run before him. But taking Mark at face value, we have to consider the possibility that they saw profit in this prophet. And Jesus evidently needed them. Why? Let's consider that as we delve into the next section of Mark. It introduces Jesus' talent for exorcism:

They went to Capernaum; and when the sabbath came, he entered the synagogue and taught. They were astounded at his teaching, for he taught them as one having authority, and not as the scribes. Just then there was in their synagogue a man with an unclean spirit, and he cried out, "What have you to do with us, Jesus of Nazareth? Have you come to destroy us? I know who you are, the Holy One of God." But Jesus rebuked him, saying, "Be silent, and come out of him!" And the unclean spirit, convulsing him and crying with a loud voice, came out of him. They were all amazed, and they kept on asking one another,

"What is this? A new teaching—with authority! He commands even the unclean spirits, and they obey him." At once his fame began to spread throughout the surrounding region of Galilee. (Mark 1:21–28, NRSV)

Ta-da! So, what role did Simon, Andrew, John, and James play in this little drama? There's no indication, but suppose Jesus were a magician . . . Well, hold on. Too much, too soon for some. Instead, let's imagine some other performer was working the Galilean circuit. There's little doubt that the Roman territory of Judaea was teeming with magicians, healers, and prophets at the time. We'll call this one . . . Brian.

Like many an illusionist, Brian needs hired help. To pull off a trick like the one described above, he would definitely need a stooge, someone to be possessed by the unclean spirit. He'd also need a plant or two, just to be sure the audience correctly interprets events. You know, someone to shout, "Say, this guy really knows his stuff!" and another to say, "Zounds! Even the demons do his bidding!" And then they'd pass the plate, pocket the take, and move on to the next hamlet.

That's Brian for you. Let's get back to Jesus. Is there any reason to think he fell into that occupation? We've slipped into tricky territory. The ground is rocky and the evidence slight. Not a single scriptural claim about Jesus can be secured by a contemporaneous source. The Bible is self-authorizing and yet, especially when it comes to Jesus, self-contradictory.

But there is at least a scrap of evidence on the question of how people saw Jesus. A bowl found in the waters off Alexandria appears to be inscribed with the words "Christ the Magician" in Greek. This wasn't high praise. Scholar Anthony LeDonne says the Greek term did not exactly conjure David Copperfield. To the literate class, anyway, magicians were rogues, con men, and troublemakers. Kind of like the Music Man rolling into town to sell a grand vision of seventy-six trombones, a hundred and ten cornets . . . and the Kingdom of God.

So, was this bowl a libel or a descriptor? Who knows? It's likely no older than the Gospels themselves, and there's nothing conclusive about it—neither the text nor its date is certain. Still, it is suggestive.

Whatever the case, Jesus wasn't the first to expel "unclean spirits," aka demons, and he certainly wasn't the last. You can watch dozens of contemporary exorcisms on YouTube, including one performed on a woman with really bad eye makeup (see https://youtu.be/zmc6lJGXe98).

Whether Jesus was regarded as a magician or a miracle worker, the act of expelling an unclean spirit remains puzzling. Was it a sign that Jesus was the Christ? If so, if exorcisms are real, then why can ordinary people perform them? What's the miracle in that? Some might claim that a priest is a vicar of Christ, drawing on the Lord's authority. Very well, but then how do you explain a Muslim exorcism, drawing on the power of Allah? (To see one performed, check: https://youtu.be/5g1KN3BRt6E.)

Okay, "Allah" refers to the same deity found in the Bible—though with a somewhat different user's manual. But then how about a Hindu exorcism? (You'll find an example of one here: https://youtu.be/Q3vFTiPCH2g?t=2m28s.) Hindus have millions of gods, but not one of them has ever turned up in the Middle East. Does that mean Hindu exorcisms are faked? Well, maybe. But if you fake one, you can fake them all. And let's face it, an exorcism is not all that hard to fake. Even a teen beauty queen can perform them. The evidence is but one more click away (see https://youtu.be/aDh3IW8vZ7o).

Which brings us back to the suggestion that Jesus was a magician. Either that, or he was a god-man, able to perform supernatural acts. Maybe so. But if that's the case, why highlight one that others can perform, or worse yet, fake? After all, if I came up to you and said, "Yo, dude, what's that coin in your ear?" brushed your lobe, and produced a shiny quarter, you wouldn't fall down and worship me, would you?

There is a postscript to this miracle. After the exorcism, the good people of Capernaum flock to Simon's house, where Jesus exorcises "silent" demons (now, there's a good trick!) and performs sundry healings. Next morning, he hits the road before dawn:

> In the morning, while it was still very dark, he got up and went out to a deserted place, and there he prayed. And Simon and his companions hunted for him. When they found him, they said to him, "Everyone is searching for you." He answered, "Let us go on to the neighboring towns, so that I may proclaim the message there also for that is what I came out to do." (Mark 1:35–38, NRSV)

Again, this is kind of suggestive. If you're the Savior, what's the rush? But if you're a flim-flam man, well, getting out of town before daybreak is standard operating procedure. Just sayin'.

21

 ## *The Clean-Hands Miracle*

The next sign is not normally reckoned a miracle, but seems worth examining all the same. It comes when Jesus is confronted by hostile Pharisees. These are religiously conservative Jews who make a big deal of observing the ritual laws. The Messiah feels they're being just a tad prissy. But the breaking point comes as a bit of a surprise.

Jesus disputes not the dress code:

> The Lord said to Moses: Speak to the Israelites, and tell them to make fringes on the corners of their garments throughout their generations and to put a blue cord on the fringe at each corner. (Numbers 15:37–38, NRSV)

Nor the grooming requirements:

> You shall not round off the hair on your temples or mar the edges of your beard. (Leviticus 19:27, NRSV)

Nor even the odd dietary laws:

> You shall not boil a kid in its mother's milk. (Deuteronomy 14:21, NRSV)

Yeah, like we were going to do that! But of all the quirky rules to oppose, Jesus chooses to take a stand against handwashing—the one thing most likely to keep his followers healthy:

Now when the Pharisees . . . gathered around him, they noticed that some of his disciples were eating with defiled hands, that is, without washing them. (For the Pharisees, and all the Jews, do not eat unless they thoroughly wash their hands, thus observing the tradition of the elders; and they do not eat anything from the market unless they wash it; and there are also many other traditions that they observe, the washing of cups, pots, and bronze kettles.) So the Pharisees and the scribes asked him, "Why do your disciples . . . eat with defiled hands?" (Mark 7:1–5, NRSV)

Jesus flies into a tirade and calls the Pharisees hypocrites for making religious offerings rather than supporting their elderly parents. "You have a fine way of rejecting the commandment of God in order to keep your tradition!" he sneers. "For Moses said, 'Honor your father and your mother'; and, 'Whoever speaks evil of father or mother must surely die.'"

Keep that last bit in mind. It's odd enough that Jesus would stand up for dirty hands, but what's miraculous is that this can be read as not contradicting what he's reported to say elsewhere.

To the much-admired Bible commentator Matthew Henry (1662–1714) the meaning of the retort to the Pharisees was perfectly clear: "One great design of Christ's coming, was, to set aside the ceremonial law which God made, and to put an end to it."

All well and good . . . and yet, in the Sermon on the Mount, Jesus says:

"Do not think that I have come to abolish the law or the prophets; I have come not to abolish but to fulfill. For truly I tell you, until heaven and earth pass away, not one letter, not one stroke of a letter, will pass from the law until all is accomplished. (Matthew 5:17–18, NRSV)

In the far more pleasing prose of the King James version, this is known as the jot and tittle passage: "For verily I say unto you, Till heaven and earth pass, one jot or one tittle shall in no wise pass from the law, till all be fulfilled."

It would take a minor miracle to reconcile either version with what we've read above, but if that weren't enough, there's also this:

Now large crowds were traveling with him; and he turned and said to them, "Whoever comes to me and does not hate father and mother, wife and children, brothers and sisters, yes, and even life itself, cannot be my disciple. (Luke 14:25–26, NRSV)

So, on the one hand Jesus declares that handwashing is so BC, and that Pharisees care more about their coffers than their mothers, and on the other he says that the laws and traditions are inviolable and that any follower of his must spurn not only their old mum and dad, but also their entire families, right down to their doe-eyed children.

Reasonable people may disagree about whether handwashing is a law, a tradition, or just good hygiene, but there is no denying that in one gospel Jesus quotes Moses saying that the penalty for dishonoring your parents is death, and in another he enjoins his followers to hate their parents.

Verily, I say unto thee, it would take a miracle to make consistent sense out of all this.

22

 Jesus Feeds Lots of People

Like any wise public speaker, Jesus made sure his audience was fed. "A full stomach makes for a happy heart," as the Spanish say. But, as we'll see, providing for multitudes can be challenging:

> In those days when there was again a great crowd without anything to eat, he called his disciples and said to them, "I have compassion for the crowd, because they have been with me now for three days and have nothing to eat. If I send them away hungry to their homes, they will faint on the way—and some of them have come from a great distance." His disciples replied, "How can one feed these people with bread here in the desert?" He asked them, "How many loaves do you have?" They said, "Seven." Then he ordered the crowd to sit down on the ground; and he took the seven loaves, and after giving thanks he broke them and gave them to his disciples to distribute; and they distributed them to the crowd. They had also a few small fish; and after blessing them, he ordered that these too should be distributed. They ate and were filled; and they took up the broken pieces left over, seven baskets full. Now there were about four thousand people. And he sent them away. (Mark 8:1–9, NRSV)

Now, this is quite a challenge, but not a physical impossibility. The account of the fish is too vague to deal with, so let's focus on the staple. According to the Museum of the University of Pennsylvania, in those times people in the region consumed about a pound of bread a day per person. The crowd attending Jesus is not going to get nearly that much. In fact, they're going to get a penny's worth each. No, not worth, actually, but little more than penny's volume. But fortunately, this is the guy who said, "Man shall not live by bread alone, but by every word that proceeds out of the mouth of God"—and he's been feeding them on the gospel truth for three days!

Let's see how the bread ration works out. Dr. Cynthia Shafer-Elliott is an associate professor of Hebrew Bible at William Jessup University in Rocklin, California. She has a fascination with the daily lives of ancient Israelites, and in pursuit of that she recreated their bread-baking practices. According to Dr. Shafer-Elliott, they baked bread in conical ovens. On an expedition to Israel, she and her students built one such oven and baked pita-style bread from scratch. It's important to note, however, that this was leavened bread. The unleavened bread, as we know, is only used ritually as a minor sacrifice to honor the trials of those who followed Moses into the wilderness.

Lacking much information about the loaves that Jesus had on hand, we're going to make some reasonable but simplifying assumptions. Although they were probably roughly cylindrical, we're going to assume that they were ten inches square and, based on some actual Roman bread that was recovered from Pompeii, two and a quarter inches thick. Now, a U.S. penny is three-quarters of an inch in diameter and 0.06 inches thick.

If Jesus breaks the bread into penny-sized wafers, we get 133 per layer. It's too much to ask him to slice bread penny-thin, so we'll allow for some depth. Let's assume that he can "break" the bread into five layers, each one nearly a half-inch thick (0.45 inches, to be precise).

Taking 133 penny-sized wafers from a ten-inch-square loaf leaves us some remainders. (The figure is actually 133.33 per loaf.) What's more, pennies are round, so the corners of each square will add to the remainders. We'll need those remainders to fill the leftover baskets.

With that in mind, we just multiply 133 x 5 layers x 7 loaves and, hey, presto! Bread for 4,655 people, with plenty of leftovers!

Miracle solved. Well, not quite. Oddly, a strikingly similar story occurs elsewhere in Mark, and then recurs in all three other gospels. Each sets the bar considerably higher.

As he went ashore, he saw a great crowd; and he had compassion for them, because they were like sheep without a shepherd; and he began to teach them many things. When it grew late, his disciples came to him and said, "This is a deserted place, and the hour is now very late; send them away so that they may go into the surrounding country and villages and buy something for themselves to eat." But he answered them, "You give them something to eat." They said to him, "Are we to go and buy two hundred denarii worth of bread, and give it to them to eat?" And he said to them, "How many loaves have you? Go and see." When they had found out, they said, "Five, and two fish." Then he ordered them to get all the people to sit down in groups on the green grass. So they sat down in groups of hundreds and of fifties. Taking the five loaves and the two fish, he looked up to heaven, and blessed and broke the loaves, and gave them to his disciples to set before the people; and he divided the two fish among them all. And all ate and were filled; and they took up twelve baskets full of broken pieces and of the fish. Those who had eaten the loaves numbered five thousand men. (Mark 6:34–44, NRSV)

Uh-oh. Now we're down to five loaves and two fish—and we've got to feed 5,000. Talk about your portion control. The account in Luke is substantially the same, as is John's—but see if you can spot one interesting difference.

When he looked up and saw a large crowd coming towards him, Jesus said to Philip, "Where are we to buy bread for these people to eat?" He said this to test him, for he himself knew what he was going to do. Philip answered him, "Six months' wages would not buy enough bread for each of them to get a little." One of his disciples, Andrew, Simon Peter's brother, said to him, "There is a boy here who has five barley loaves and two fish. But what are they among so many people?" Jesus said, "Make the people sit down." Now there was a great deal of grass in the place; so they sat down, about five thousand in all. Then Jesus took the loaves, and when he had given thanks, he distributed them to those who were seated; so also the fish, as much as they wanted. When they were satisfied, he told his disciples, "Gather up the fragments left over, so that nothing may be lost." So they gathered them up, and from the fragments of the five barley loaves, left by those who had eaten, they filled twelve baskets. When the people saw the sign that he had done, they began to say, "This is indeed the prophet who is to come into the world." (John 6:5–15, NRSV)

So, at first Jesus wants to buy food for the masses, but when he's told the price is too steep, what happens? Apparently, the disciples, um, appropriate the food from some poor kid who was probably on his way home from doing the shopping his ma sent him to do. What's she gonna say when he tells her that Jesus took the family's groceries to feed 5,000 loafers sitting in a field?

Theologians write that the boy shared his food, but there's no textual evidence of that. Perhaps he got paid for it, but again, there's no evidence. At least, we can hope he got a portion of the bread that was going around. For evidence, we turn to the final gospel account of the feeding of the multitudes.

In Matthew, the crowd is even larger! Turns out all this time, in true biblical fashion, we've been ignoring the women and children.

Now when Jesus heard this, he withdrew from there in a boat to a deserted place by himself. But when the crowds heard it, they followed him on foot from the towns. When he went ashore, he saw a great crowd; and he had compassion for them and cured their sick. When it was evening, the disciples came to him and said, "This is a deserted place, and the hour is now late; send the crowds away so that they may go into the villages and buy food for themselves." Jesus said to them, "They need not go away; you give them something to eat." They replied, "We have nothing here but five loaves and two fish." And he said, "Bring them here to me." Then he ordered the crowds to sit down on the grass. Taking the five loaves and the two fish, he looked up to heaven, and blessed and broke the loaves, and gave them to the disciples, and the disciples gave them to the crowds. And all ate and were filled; and they took up what was left over of the broken pieces, twelve baskets full. And those who ate were about five thousand men, besides women and children. (Matthew 14:13–21, NRSV)

Okay, so now we've got five loaves for . . . what? Ten thousand people? Twenty thousand? Folks had a lot of kids in those days. Many more could be fed if the portions were reduced to the size of a modern communion wafer—well over 20,000 by my estimate. But dividing bread loaves into wafers would have required industrial production methods. If Jesus had had modern slicing and stamping equipment, surely someone would have mentioned it. He'd be better off using a *Star Trek*–style replicator. (If Jesus can use technology from our era, why not from the future?)

But hold on. Nature teems with actual replicators. If five loaves became 5,000 by natural processes, it had to be via one of these: Crystals are replicators. But they only induce like materials to get organized in pleasing symmetrical formations. Another replicator, the one that happens to make food for us, is the living cell. Bacteria are expert replicators. Provided it doesn't experience mutation or exchange genetic material with other cells, a bacterium can divide to become two identical bacteria. And many species of bacteria can do this with astonishing speed. Under ideal conditions, some bacteria can divide four or five times an hour. Again, ideally, this means growth at an amazing clip.

Suppose that loaves could do something similar? In an hour one loaf would become 16 loaves, but in just two and a half hours, one loaf would have multiplied into 1,024 loaves. Starting with five loaves, then, exponential growth would mean 5,120 loaves to share among the masses. That ought to do the trick.

Unfortunately, that would require at least 5,000 pounds of raw material for the replicants to feed on. What's more bread is not made from bacteria (yuck!), but from wholesome, slow-growing wheat. Even the living yeast that gives rise to it cannot reproduce once it's baked. Alas, replication doesn't yield a realistic answer.

For now, consider the likelihood of another explanation: people exaggerate. The stories we tell enhance our reputations. That's why the fish that got away has to be bigger than any in the catch. Sometimes our exaggerations have additional motives. In Philadelphia, where I grew up, the public transit authority was plagued by an interesting form of exaggeration: whenever there was an accident involving one of its buses or trolleys, many more people sued for compensation than could possibly have fit on the vehicle. Talk about feeding the five thousand!

We don't need to impute base motives to early Christians to see how likely it is that their stories would be exaggerated as they went from mouth to mouth. Remember, even the most conservative estimate of the earliest gospel has it written more than a decade after Jesus' death. The scholarly consensus puts the earliest date at 70 CE. That's an ocean of time for an oral tradition to grow.

Crowd counts are notoriously difficult, even today. Just ask President Trump about his inauguration. The gospel detail of sitting down in "groups of hundreds and fifties" is an oddity that raises more questions than it answers. Did the people count off? Could a people who were largely without schooling even do so? Who checked to see how many per group?

Other crowd counts in the Bible give us little confidence in their accuracy. Remember, Moses supposedly led 1.2 million people into the Sinai Peninsula, where they somehow fed and watered themselves and their cattle for forty years—leaving not a trace behind.

So, in general we have reasons to be skeptical about size of the multitude, but there's a specific reason as well: the two accounts in Mark. These make the case for exaggeration solid. After all, if miracles are meant to be convincers, only the first one is needed. Feeding 5,000 men? That's a whopper.

But as we've seen, feeding 4,000 with seven loaves only requires manual dexterity. A pinch here, a pinch there, and everybody nibbles. Narratively, it comes as an anticlimax, following the feeding of the 5,000. So, what's it doing there? Jesus addresses dozens of crowds in the Gospels. Why feed only two groups? And why's the menu the same? Surely, this suggests a common origin. It is likely a single story that evolved into two species of the same coinage. But, like heads and tails, there's an essential difference between them. One is a (barely) realistic depiction of what may have happened; the other a supernatural event or an exaggeration intended to convince readers of the divine nature of Jesus.

23

 The Transfiguration of Jesus

In Matthew, there's an odd little episode that, in modern political terms, amounts to an endorsement. In Christianity, however, it is known as the Transfiguration, a scene that has provided gainful employment to generations of classical artists. Here's the text:

> Jesus took with him Peter and James and his brother John and led them up a high mountain, by themselves. And he was transfigured before them, and his face shone like the sun, and his clothes became dazzling white. Suddenly there appeared to them Moses and Elijah, talking with him. Then Peter said to Jesus, "Lord, it is good for us to be here; if you wish, I will make three dwellings here, one for you, one for Moses, and one for Elijah." While he was still speaking, suddenly a bright cloud overshadowed them, and from the cloud a voice said, "This is my Son, the Beloved; with him I am well pleased; listen to him!" When the disciples heard this, they fell to the ground and were overcome by fear. But Jesus came and touched them, saying, "Get up and do not be afraid." And when they looked up, they saw no one except Jesus himself alone. (Matthew 17:1–8, NRSV)

Cool. But is it a miracle? The answer may depend on whether you accept Clarke's Third Law: *Any sufficiently advanced technology is indistinguishable from magic.* To illustrate the point, first consider the Italian painter Titian's version of the Transfiguration rendered in oils. And then think about what a modern equivalent might look like as a holographic projection. You can find a fine example of a dazzling holography show in Dubai on YouTube at https://youtu.be/3d7sQfIBAwk. If you watch the video, you'll notice that children on the stage can see the projection of tigers, dolphins, and storm clouds, just as those in the audience can. What would the Apostles have made of that?

It may be amazing, it's certainly way cool, but no one from today's world would believe for an instant that a holograph is, literally, a miracle. It's actually a sophisticated projection system that makes use of thin Mylar screens stretched across the stage and overhead to create live, lifelike images seemingly floating in thin air. For an explanation of how the system works, you can watch another YouTube video here: https://youtu.be/pSICZ_7hpho. But we don't have to understand the details to be sure that this is not magic. We just have to have confidence that technology must operate within the laws of physics.

Ancient peoples could have no such confidence. For one thing, they had barely any technology, and certainly no electronics. But equally important, they had only a rough, intuitive sense of the laws of physics. If there's one thing that science has demonstrated over the last few centuries, it's that intuition doesn't get you very far in physics. A rock may feel absolutely solid in our hands, but it's mostly space, permeated only by electromagnetic force fields. In fact, you can't touch it at all— the electrons in your hands repel the electrons in the rock and they fail to pass through each other. That's as close as "touch" normally comes in the subatomic world.

So, could God and Jesus have pulled off a holographic projection stunt to amaze Peter, James, and John? Your first reaction may be to scoff at the notion, but hold on. Once again, it's technomancy time.

If you accept the claim that God is all knowing, that must mean he knows about all the technology that will ever be. Now, I should note that the idea of being omniscient is philosophically paradoxical (how can God know himself?) and physically impossible (information, we now know, comes in physical bits, and every bit used to represent something else is a bit that cannot represent itself. Still, if you accept omniscience on faith, then it follows that God knows all about technology, from portable solar power generators to digital projectors to Mylar films to molecular 3-D printers of the future.

With all that in mind, this passage takes on new significance: Jesus leads the boys into the mountains, away from prying eyes, to a place where all the equipment has been set up out of sight. The sequences, as described, would take only a few moments, and when the disciples prostrate themselves in fear, is it really too much to imagine Jesus making the director's "Cut!" sign? At any rate, when the boys look up, there's nothing to see but Jesus himself.

Now, some may object that mixing God, Jesus, and electronic technology makes a hash of religion and science. Very well. All we have to do is replace supernatural beings with mischievous time travelers from the future or aliens with a sense of humor. All natural ingredients, same result.

24

 Revival Show

The Gospels report numerous instances of Jesus healing others. In a case of what appears to be excessive earwax buildup, the Lord employs warm spit to unclog a deaf man's ears. But where Jesus really strives to set himself apart from the other spiritual healers is in the raising of the dead. Once again, we begin with the earliest of the Gospels, Mark. Jesus hops a ferry across the Galilee. On the opposite shore he meets one Jairus.

> When Jesus had crossed again in the boat to the other side, a great crowd gathered around him; and he was by the sea. Then one of the leaders of the synagogue named Jairus came and, when he saw him, fell at his feet and begged him repeatedly, "My little daughter is at the point of death. Come and lay your hands on her, so that she may be made well, and live." So he went with him. (Mark 5:21–24, NRSV)

On the way, however, Jesus is delayed by a lady who has . . . well, female troubles.

> Now there was a woman who had been suffering from hemorrhages for twelve years. She had endured much under many physicians, and had spent all that she had; and she was no better, but rather grew worse. (Mark 5:25–26, NRSV)

Now, I'm no doctor, but from this description it seems likely the woman was suffering from a fibroid in her uterus. (Apart from nosebleeds, other kinds of hemorrhage would likely have long ago killed her.) Uterine fibroids are noncancerous growths that commonly result in abnormal vaginal bleeding. Today's medical treatments range from birth control pills to surgery. But having spent all her savings on ancient Hebrew quacks, this poor lady has but one last gambit to play:

> She had heard about Jesus, and came up behind him in the crowd and touched his cloak, for she said, "If I but touch his clothes, I will be made well." Immediately her hemorrhage stopped; and she felt in her body that she was healed of her disease. (Mark 5:26–29, NRSV)

That's great, but apparently the conservation laws apply even to supernatural energy, for Jesus can feel the drain on his batteries:

> Immediately aware that power had gone forth from him, Jesus turned about in the crowd and said, "Who touched my clothes?" And his disciples said to him, "You see the crowd pressing in on you; how can you say, 'Who touched me?' " He looked all around to see who had done it. But the woman, knowing what had happened to her, came in fear and trembling, fell down before him, and told him the whole truth. He said to her, "Daughter, your faith has made you well; go in peace, and be healed of your disease." (Mark 5:30–34, NRSV)

Unfortunately, all this dallying over a quick and inexpensive cure for fibroids has come at a steep price. Villagers run up with the sad news: the daughter of Jairus has died. But Jesus isn't about to let Death stand in his way:

> When they came to the house of the leader of the synagogue, he saw a commotion, people weeping and wailing loudly. When he had entered, he said to them, "Why do you make a commotion and weep? The child is not dead but sleeping." And they laughed at him. Then he put them all outside, and took the child's father and mother and those who were with him, and went in where the child was. He took her by the hand and said to her, "Talitha cum," which means, "Little girl, get up!" And immediately the girl got up and began to walk about (she was

twelve years of age). At this they were overcome with amazement. He strictly ordered them that no one should know this, and told them to give her something to eat. (Mark 5:38–43, NRSV)

Now, this is rather odd. Jesus *tells* us that the girl is not dead. Is he speaking metaphorically? Evidently not. Yet, those who believe she is dead take time out from their grief *to laugh at him*. Really? One moment they are "weeping and wailing loudly," and the next they are all like, "Ah, ha, ha, ha. What a silly little man."

Am I the only one who finds this hard to credit?

Anyway, to prove them wrong, Jesus clears the room, takes the girl by the hand, and tells her to wake up—and she does. Wonderful? Sure. Touching? Absolutely. Incredible? Not really. She may have passed out. She may have lapsed into a coma. At any rate, we have no reason to think Jesus was wrong when he said, "The child is not dead but sleeping."

But then comes another odd twist: Earlier that day the Savior told the guy he cured of possession, "Go home to your friends, and tell them how much the Lord has done for you." Now he tells everyone to clam up. "He strictly ordered them that no one should know this . . ."

And yet we do. Some snitch must be roasting in hell.

As secretive as Jesus was about reviving the daughter of Jairus, he's not at all shy about performing a highly public reanimation in a town called Nain.

As he approached the gate of the town, a man who had died was being carried out. He was his mother's only son, and she was a widow; and with her was a large crowd from the town. When the Lord saw her, he had compassion for her and said to her, "Do not weep." Then he came forward and touched the bier, and the bearers stood still. And he said, "Young man, I say to you, rise!" The dead man sat up and began to speak, and Jesus gave him to his mother. Fear seized all of them; and they glorified God, saying, "A great prophet has risen among us!" and "God has looked favorably on his people!" This word about him spread throughout Judea and all the surrounding country. (Luke 7:11–17, NRSV)

Once again, if we take the story at face value and consider the range of natural possibilities, it is just barely explicable. It is possible for a person to appear dead, even to experts, and then revive.

At Saint Joseph's Hospital in Syracuse, New York, for example, this happened: 41-year-old Colleen Burns was declared dead of cardiac arrest following a drug overdose. Her family agreed to donate her organs. Just before surgeons were about to cut her open and scoop them out, she woke up. Two weeks later, she went home just fine.

Of course, this was not an everyday event. Authorities later fined the hospital $22,000 for errors in the case. Nevertheless, it proves the point: death can be a bit of a prankster.

Until the late twentieth century, the medical standard for determining death was absence of heartbeat. However, Jewish law begs to differ. According to physician Abraham Steinberg of Hebrew University, the codified tradition holds that an absence of breath marks the end of life. As noted earlier, the great Jewish philosopher and physician Moses Maimonides recommended holding a mirror under a patient's nostrils to detect breath—but that was ten centuries after Jesus. Who knows what may have been the standard for determining death in first-century Nain? Actually, we can be pretty sure no standard existed, because there certainly was no medical doctor with a mirror in his satchel, and there was no clear conceptual line between sleep and death. To speak of a (presumably) dead person as asleep is not merely a rhetorical flourish. Paul, who lived at the same time but never met Jesus in the flesh, repeatedly conflates sleep and death in his letters. For example, if we look at an accurate translation from the Greek, we find that Paul writes:

> [W]e do not want you to be uninformed, brothers and sisters, about those who have *fallen asleep*, so that you may not grieve as others do who have no hope. For since we believe that Jesus *died* and rose again, even so, through Jesus, God will bring with him those who have *fallen asleep*. [Emphasis added.] (1 Thess. 4:13–14)

So, when we hear that the widow of Nain's son was dead, we have to wonder. Is it possible that rather than bringing him back from the dead, Jesus saved him from death? Maybe the poor kid was in a coma. That word, by the way? It comes from ancient Greek and means "deep sleep."

Wakefulness, on the other hand, is unambiguous. But how it works is something we're just beginning to understand. Dr. Clifford B. Saper, chair of neurology at Harvard Medical School, tells us that wakefulness is maintained by a clump of cells in the brainstem.

> These nerve cells use excitatory neurotransmitters such as acetylcholine, norepinephrine, dopamine, and glutamate, to turn on cell groups in the upper part of the brain, called the forebrain. The ultimate target of this arousal system is the cerebral cortex, the part of the brain responsible for perception, thought, and behavior.

If something like a blow to the back of the head disrupts the flow of excitatory neurotransmitters, consciousness winks out. Depending on the severity of the blow, the flow may soon restore itself, or consciousness may be suppressed for years and possibly forever.

Apart from physical trauma, viruses are a fairly common cause of disruption. Some, such as West Nile virus, can get past the brain's defenses and cause havoc. The name for such attacks is encephalitis. When it strikes, the possibilities range from a bad headache to coma, and even death. In the absence of modern drugs, the outcome depends mainly on the virulence of the invader versus the strength of the body's immune system.

So, is it possible that the widow of Nain's son lay in an encephalitic coma? Quite possible. Could Jesus have revived him? That's problematic. The main difference between a coma and sleep is that you cannot wake a person from a coma by calling their name or shaking them. If coma were the cause, it would have needed exquisite timing for Jesus to take credit for the young man's revival.

What else might explain this? The kid could have been diabetic. It's only in the last century that diabetes has been treatable, but there's nothing new about the condition. The ancient Egyptian physician Hesy-Ra described it way back in 1552 BCE. He knew nothing about insulin or resistance to its uptake, but he could note the symptoms.

We now know that an inability to produce or use insulin can cause the brain to run low on fuel. Like the rest of the body, the brain runs on glucose, a sugar that's carried in our bloodstream. When it can't get enough (or sometimes, when the tank is flooded), the brain shuts down consciousness to save energy. If untreated, diabetic coma *can* lead to death, but even in the absence of treatment, spontaneous recovery within a few hours is common. Maybe Jesus happened upon a boy recovering from diabetic shock.

There is yet another possibility. The young scalawag could simply have been on a bender. A person who has passed out after getting wretchedly drunk can be mistaken for dead. Often, they'll snore loudly, but not always. The expression "dead drunk" has its roots in reality. Could it be, then, that in reviving Nain's No. 1 bad lad, Jesus cast out . . . demon rum?

25

 Lucky Lazarus

Let us not forget that the types of miraculous revivals attributed to Jesus had also been attributed to others. According to the Bible, the prophets Elijah and Elisha both performed the seemingly impossible act. But if reawakening the dead was meant to identify Jesus as the Son of God, why would mere prophets also have had the power?

Possibly, that very question came up among skeptical Jews confronting followers of Jesus in the decades after his death. After all the new "Christians" were asking their fellow Jews to accept a truly radical idea: not just that the long-dead Jesus was the Messiah, but also that "Messiah" meant something way beyond what the Jews had long believed.

So perhaps that's why the final and most famous revival story ups the ante. In the Gospel of John, Lazarus becomes a pivotal figure in the life and death of Jesus himself.

The author known as John makes the revival of Lazarus the capstone on a colonnade of miracles. Uniquely among the Gospels, it proves to be the act that seals the Son of God's fate. Kudos to John for a clever plot device: in bringing back Lazarus, Jesus puts on a show that prefigures his own death and resurrection. And a show it most certainly is, one that opens with a surprising reaction to dreadful news.

> Now a certain man was ill, Lazarus of Bethany, the village of Mary and her sister Martha. Mary was the one who anointed the Lord with perfume and wiped his feet with her hair; her brother Lazarus was ill.

> So the sisters sent a message to Jesus, "Lord, he whom you love is ill."
> But when Jesus heard it, he said, "This illness does not lead to death;
> rather it is for God's glory, so that the Son of God may be glorified
> through it." Accordingly, though Jesus loved Martha and her sister and
> Lazarus, after having heard that Lazarus was ill, he stayed two days
> longer in the place where he was. (John 11:1–6, NRSV)

Jesus has come a long way from his private, no-one-must-know revival of Jairus's daughter to this deliberate "let him ripen a few days" demonstration of his powers. This may seem heartless, but remember, "it is for God's glory." Besides, Jesus is in a wild, metaphor-spinning mood.

> Then . . . he said to the disciples, "Let us go to Judea again." The
> disciples said to him, "Rabbi, the Jews were just now trying to stone
> you, and are you going there again?" Jesus answered, "Are there not
> twelve hours of daylight? Those who walk during the day do not
> stumble, because they see the light of this world. But those who walk
> at night stumble, because the light is not in them." After saying this,
> he told them, "Our friend Lazarus has fallen asleep, but I am going
> there to awaken him." The disciples said to him, "Lord, if he has fallen
> asleep, he will be all right." Jesus, however, had been speaking about
> his death, but they thought that he was referring merely to sleep. Then
> Jesus told them plainly, "Lazarus is dead. For your sake I am glad I was
> not there, so that you may believe." (John 11:7–15, NRSV)

After this "nothing up my sleeves" speech, Jesus eventually pulls up stakes and leads his disciples to nearby Bethany. Even then, it's a leisurely trip. By the time they arrive, his friend has long been dead. Martha upbraids the Lord for his tardiness.

> When Jesus arrived, he found that Lazarus had already been in the
> tomb four days. Now Bethany was near Jerusalem, some two miles
> away, and many of the Jews had come to Martha and Mary to console
> them about their brother. When Martha heard that Jesus was coming,
> she went and met him, while Mary stayed at home. Martha said to
> Jesus, "Lord, if you had been here, my brother would not have died."
> (John 11:18–21, NRSV)

Jesus then makes a bold declaration, which runs smack into one of the fundamental contradictions in the Bible. Do people who die go straight to heaven (or hell) as spirits? Or do they remain their corporeal selves, moldering in the grave, and tapping their long fingernails on the lid while they await physical resurrection in the Last Days? Jesus hints that it's the latter. Better put a deck of cards in that coffin. It's gonna be a long wait.

> Jesus said to her, "Your brother will rise again." Martha said to him, "I know that he will rise again in the resurrection on the last day." Jesus said to her, "I am the resurrection and the life. Those who believe in me, even though they die, will live, and everyone who lives and believes in me will never die. Do you believe this?" She said to him, "Yes, Lord, I believe that you are the Messiah, the Son of God, the one coming into the world." (John 11:23–27, NRSV)

Soooo, Laz just has to wait his turn, along with all the other believers, right? Nothing to see here, folks, move along. Well, not quite done. Martha believes, but maybe Lazarus hadn't gotten the word. Anyway, there's still a crowd of friends and relations milling about, and they are none too happy with Jesus. So, it's on to Act 2.

> When she had said this, she went back and called her sister Mary, and told her privately, "The Teacher is here and is calling for you." And when she heard it, she got up quickly and went to him. Now Jesus had not yet come to the village, but was still at the place where Martha had met him. The Jews who were with her in the house, consoling her, saw Mary get up quickly and go out. They followed her because they thought that she was going to the tomb to weep there. When Mary came where Jesus was and saw him, she knelt at his feet and said to him, "Lord, if you had been here, my brother would not have died." When Jesus saw her weeping, and the Jews who came with her also weeping, he was greatly disturbed in spirit and deeply moved. He said, "Where have you laid him?" They said to him, "Lord, come and see." Jesus began to weep. So the Jews said, "See how he loved him!" But some of them said, "Could not he who opened the eyes of the blind man have kept this man from dying?" (John 11:28–37, NRSV)

This is a famous passage, source of the common expression, "Jesus wept." But to the literal-minded reader (as we are enjoined to be), it is a puzzle. Jesus has said from the start that he is putting on a demonstration, and he surely knows what's going to happen next. So, why the tears? Yes, they show his compassion, but unless there is genuine feeling behind them, then that compassion is merely for show. Could Jesus be toying with them?

Surely, the Son of Man knows that letting a man languish, die, and then decay for a couple of days, all so that he can show off his mighty powers, is a mite unethical. Does Jesus have a choice? If so, what's he—and what are we—to make of his own fate? Who's running this show? No time to address those questions now; there's a corpse to resurrect.

> Then Jesus, again greatly disturbed, came to the tomb. It was a cave, and a stone was lying against it. Jesus said, "Take away the stone."

Regrettably, Laz's remains have not aged like a rib eye steak. There's a bit of a pong in the air. However unpleasant, though, the odor is part of an overall marketing plan.

> Martha, the sister of the dead man, said to him, "Lord, already there is a stench because he has been dead four days." Jesus said to her, "Did I not tell you that if you believed, you would see the glory of God?" So they took away the stone. And Jesus looked upward and said, "Father, I thank you for having heard me. I knew that you always hear me, but I have said this for the sake of the crowd standing here, so that they may believe that you sent me." When he had said this, he cried with a loud voice, "Lazarus, come out!" The dead man came out, his hands and feet bound with strips of cloth, and his face wrapped in a cloth. Jesus said to them, "Unbind him, and let him go." (John 11:38–44, NRSV)

Is such a thing possible? Absent deceit, it is extremely improbable—but not impossible. We'll defer the question of whether anyone can recover from actual death. It remains medically plausible that Lazarus conked out for a few days, developed a stench, then recovered.

One scenario: Lazarus might have had GAS. Settle down. It's not what you're thinking. Martha's brother might have had the misfortune to be infected by a Group A Streptococcus bug. *Streptococcus pyogenes*, in particular, is a nasty little bacterium that occasionally makes its way into our bodies. The most familiar route is through the vulnerable tissues of the esophagus, where it gives rise to strep throat. But a puncture wound can also carry it from the surface of the skin, where it likes to hang out, into subcutaneous tissues and the bloodstream. There, GAS can wreak havoc.

The toxins released in a GAS infection can necrotize (that is, kill) the tissues around the point of entry. Rotting tissues soon produce a stink. Meantime, the bloodstream infection causes toxic shock syndrome, whose symptoms include rapid fall in blood pressure, leading to unconsciousness. Together, to the untrained eye, these could certainly mimic death. And in many instances, death would indeed be hovering nearby. Untreated, toxic shock syndrome often proves fatal . . . but not always. In some cases. the immune system is able to adapt quickly enough to extinguish the infection within days. Lazarus may have been one such lucky sod.

If so, he need not have awakened at the very moment Jesus called to him. His tomb, we are told, was covered with a large rock. Laz would have been weak and disoriented when he returned to consciousness— and no doubt surprised to find himself wrapped like a mummy and entombed. He would have surely been weak. But when the rock rolled away and Jesus called to him, the most natural response would be to try to rise and get the hell out of that stinking cave.

Jesus may have convinced those who were on hand of his godly powers, but evidently the story was not good enough for those in high places. That powerful sect known as the Pharisees had plenty of reason to worry about this upstart rabbi, but if they accepted his miracles at face value, they surely would not have responded to the raising of Lazarus in the following way.

> Many of the Jews therefore, who had come with Mary and had seen what Jesus did, believed in him. But some of them went to the Pharisees and told them what he had done. So the chief priests and the Pharisees called a meeting of the council, and said, "What are we to do? This

man is performing many signs. If we let him go on like this, everyone will believe in him, and the Romans will come and destroy both our holy place and our nation." (John 11:45–49, NRSV)

The author of John writes this after 70 CE, when the Romans did indeed sack the Temple and destroy Jerusalem. The connection with Jesus, crucified more than a generation earlier, is hazy to say the least. But the very idea that these Pharisees would plot against the Son of God, *if they recognized him as such*, is absurd. Only Pharaoh could be that dumb. Oh, but come to think of it, God was manipulating that poor Egyptian sap like a pawn on a chessboard. And, lo and behold, he's at it again.

But one of them, Caiaphas, who was high priest that year, said to them, "You know nothing at all! You do not understand that it is better for you to have one man die for the people than to have the whole nation destroyed." He did not say this on his own, but being high priest that year he prophesied that Jesus was about to die for the nation, and not for the nation only, but to gather into one the dispersed children of God. So from that day on they planned to put him to death. (John 11:49–53, NRSV)

And that, dear reader, foreshadows the biggest miracle of them all. But before we arrive at the Resurrection, there are more fun miracles yet to examine. Get your water wings on!

26

 Walking on Water

Missed the boat? No problem, if you're the Son of God. Just stroll over the Sea of Galilee and climb aboard.

This is surely the most famous of all miracles attributed to Jesus. So famous that it has become a sardonic phrase about pompous or sanctimonious leaders: "Look at him. He thinks he walks on water." It also pops up in the title and refrain of a really sappy Randy Travis song about his grandfather: "He Walked on Water." All of which is kind of funny when you consider that water walking is an everyday feat for bugs. At least, bugs belonging to the family called Gerridae.

What we learn from them is that walking on water is easy if your body is (a) water repellent, (b) small, and (c) spread out sufficiently that it is too light to break through the surface of the water. That's possible thanks to surface tension, brought about by water molecules liberally engaging in group hugs.

There are three accounts of Jesus' water walking in the Gospels, beginning with Mark's:

> Immediately he made his disciples get into the boat and go on ahead to the other side, to Bethsaida, while he dismissed the crowd. After saying farewell to them, he went up on the mountain to pray. When evening came, the boat was out on the sea, and he was alone on the land. When he saw that they were straining at the oars against an adverse wind, he came towards them early in the morning, walking on the sea. He intended to pass them by. But when they saw him walking on

the sea, they thought it was a ghost and cried out; for they all saw him and were terrified. But immediately he spoke to them and said, "Take heart, it is I; do not be afraid." Then he got into the boat with them and the wind ceased. And they were utterly astounded, for they did not understand about the loaves, but their hearts were hardened. (Mark 6:45–52, NRSV)

Matthew's account is almost the same, except that he gets rid of that curious mention of the loaves near the end. Now, some Bible commentators claim this is a reference to the previous miracle of feeding a crowd by making a couple of bread loaves go a long, long way. But any *Mythbusters* fan will know what's really meant here. Jesus was trying out Ninja floatation shoes! Unfortunately, as cohost Adam Savage proved, walking on water with "loaves" on your feet is nigh unto impossible. Just staying upright for more than a few moments amounts to a minor miracle; any attempt to walk tips you over.

Well, there's your problem. Maybe Jesus used his tunic as a sail. Then, the disciples, not understanding that the loaves were for floating, thought Jesus was offering them a soggy, tasteless snack—with his feet! No wonder their hearts were hardened.

Matthew deletes this awkward interlude, and then adds one or two embellishments of his own:

Immediately he made the disciples get into the boat and go on ahead to the other side, while he dismissed the crowds. And after he had dismissed the crowds, he went up the mountain by himself to pray. When evening came, he was there alone, but by this time the boat, battered by the waves, was far from the land, for the wind was against them. And early in the morning he came walking toward them on the sea. But when the disciples saw him walking on the sea, they were terrified, saying, "It is a ghost!" And they cried out in fear. But immediately Jesus spoke to them and said, "Take heart, it is I; do not be afraid."

Peter answered him, "Lord, if it is you, command me to come to you on the water." He said, "Come." So Peter got out of the boat, started walking on the water, and came toward Jesus. But when he noticed the strong wind, he became frightened, and beginning to sink, he cried out, "Lord, save me!" Jesus immediately reached out his hand and caught him, saying to him, "You of little faith, why did you doubt?" When they got into the boat, the wind ceased. And those in the boat worshiped him, saying, "Truly you are the Son of God." (Matthew 14:22–33, NRSV)

The account in John is quite short, but includes some key details.

When evening came, his disciples went down to the sea, got into a boat, and started across the sea to Capernaum. It was now dark, and Jesus had not yet come to them. The sea became rough because a strong wind was blowing. When they had rowed about three or four miles, they saw Jesus walking on the sea and coming near the boat, and they were terrified. But he said to them, "It is I; do not be afraid." Then they wanted to take him into the boat, and immediately the boat reached the land toward which they were going. (John 6:16–21, NRSV)

So, three fairly similar accounts have Jesus walking across the Sea of Galilee under strong winds. How can we account for this? If you believe everything you see on YouTube, you might think that Jesus ran on water. "You have to believe in it," say the self-styled Liquid Mountaineers in a video that garnered them over 14 million views. Their video purports to show them running on water in water-repellent shoes. But then, if you believe that, you may also believe one of the hundreds of YouTube videos that claim the Second Coming is a done deal, making the rest of this book, your bank balance, your golf handicap, and pretty much everything else totally irrelevant. Pardon me, I have to go repent.

* * *

Okay, I'm back, but apparently Jesus is not, so let's get on with it.

Alas, the shoe does not fit. Even the best of the alleged "Liquid Mountaineers" only stays afloat for half-a-dozen strides, presumably the length of the subsurface ramp he's running on. Yes, it's a fraud. With the assistance of an Olympic sprinter, *Mythbusters* demonstrated that it cannot work. People are not "Jesus" lizards.

A more ingenious, less disingenuous explanation has been offered by a team of American and Israeli scientists: maybe Jesus walked on ice. If water in its liquid form has a tendency to stick together, once H_2O turns to ice it really resists being pushed around. As water cools, its molecules stop moving about so energetically, and this allows more and more of its electrons to loop into each other's paths—joining hands, so to speak—to form a solid.

But what would ice be doing in the Middle East? Well, it seems that when you have freshwater lakes with warm saltwater springs running into them, conditions that certainly obtained in the ancient Sea of Galilee, you can occasionally get a phenomenon known as "springs ice." These are ice sheets that form right alongside open water. In a *National Geographic* article, lead scientist Doron Nof puts it this way: "A rare set of weather events may have combined to create a slab of ice about 4 to 6 inches [10 to 15 centimeters] thick on the lake, [making it] able to support a person's weight. . . . And our models show that there was also a cold snap at that time, which lasted a few days and drastically lowered the temperature."

That's where timing comes in. If the lake were to partially freeze, it would surely take all night to do so. Therein lies the significant difference between the Mark-'n'-Matt account and John's. Of course, John is not all that specific; he only says it was dark. Perhaps we could read this as saying that Jesus walked out in the predawn hours. At any rate, there's a difference between the gospels, so we are free to explore the one that best fits the scientific case. Let's start with Mark, which presents the earliest version of the story.

It's a pretty simple tale: Jesus sees that his boys in the boat are straining against the wind and making no progress. Jesus walks out toward them. An odd detail pops in: "He intended to pass them by." Was he going to put his thumbs in his ears, waggle his fingers, and say, "Nyah, nyah, slowpokes," as he strode past? Well, let's be charitable and assume that Mark means that Jesus had intended to walk across the entire Sea of Galilee. When his disciples cry out in fear, taking him for a ghost, Jesus clambers aboard their boat, astonishes one and all with his "loaves," and, lo, their hearts are hardened. Whatever that last phrase means, we can safely assume it does not refer to myocardial calcification. One less research project for your poor, overworked author.

Back to the skating rink. If you accept that ice might have formed along the shore to an extent and thickness that a man could walk out on it for a hundred yards or so, then we have a plausible explanation. Case closed? Not exactly. Nof and his colleagues succeed in showing that ice can form in Israel, but scientific critics say the strong winds mentioned in all the gospel accounts probably would have disrupted the process. Who's right? Who knows . . .

Besides, there are the other gospel accounts. Let's take a closer look at the challenges they pose. Matthew adds the humorous touch. As Jesus approaches Peter asks permission to get out of the boat and join him, clambers out and for a moment walks on water, just like Jesus. But then, like the cartoon character Wile E. Coyote running off the edge of a cliff, Peter notices what he's doing, and . . . whoop whoop . . . Ker-splash!

Ah, the Bible. It's comedy gold.

Matthew's version has Jesus chiding Peter for letting doubts creep in. Far from supernatural, this is a very human moment, as anyone who has ever tried to walk a tightrope knows. In a situation like that, the last thing you want to do is consider what you are doing. But this only makes sense if Peter was actually responsible for his own flotation. Maybe the poor sap stepped onto a breakaway chunk of ice, managing to stay upright for only a few moments.

Alternatively, Jesus, extending his magical floating powers to Peter, withdraws them the moment Peter takes flight. But that's just mean. If the Son of Man wanted to teach Peter to have confidence in the Lord's grace, playing the prankster seems like poor pedagogy.

There's another difficulty with the later gospel accounts. Matthew puts the boat "far" at sea, while John gets more specific and says, "Three or four miles."

Now, that's a problem. Astute readers may wonder how the Bible, written thousands of years ago, could make use of miles. It doesn't. Instead, modern translators have taken the Greek term *stadia* and converted it into miles. But by any measure there's just no way Jesus can see his disciples struggling with their oars at that distance. It's not only that he'd need Superman's telescopic vision. However good your sight, you cannot peer over the horizon. On level ground, a person six feet tall can at most see 3 miles into the distance. But let's face it: Semites are not a tall people. We're not talking Masai here. Even today, the average height for Middle Eastern men is around 5'7" (171 cm). Back then, people were definitely shorter. We know that ancient Egyptian royalty topped out at about 5'2". Was Jesus a great exception to the rule? If he'd had NBA potential, you'd think it would be mentioned somewhere in the Gospels.

If he were anywhere close to average height for his time, he'd not have been more than 5'5". That would have limited his viewing range to less than the minimum 3 miles that John gives. And even if he stood on a Roman soapbox, there's a problem: the offing. That is a distant zone in the ocean or a large lake in which a ship is visible but only just. Next time you're standing on a shoreline, look out at the horizon. You'll notice that the offing appears to be of a different color. That's due to foreshortening. But if there are any boats visible in it, you may also notice that they appear to be distorted, and not in a nice, stable way. Images in the offing bob and twist into ever-changing shapes, like a boiling cloud. This is because the air over the surface of a body of water differs in temperature: the closer to shore the warmer (or cooler) it is, depending on time of day. It also differs from the temperature of the air over land, so the light passing through gets buffeted as it passes through differing layers of air. (You can see something similar on a long, straight highway.)

So, even if Jesus had the long view, it's highly unlikely he could have made out his shipmates struggling against the winds. On top of all that, let's not forget, it's either dark or early morning. Did the Lord have a searchlight?

If not, we'd have to believe that Jesus set out in the dark to cross the Sea of Galilee and just happened to come across his buddies. Or maybe the tale—and the reported distance—stretched in the telling from Mark to Matthew to John. Certainly, the in-shore version in Mark makes more sense, and has the additional virtue of allowing for the springs ice explanation—however thin that may be.

One last consideration: like many Bible stories, this is only considered miraculous because of the poor state of technology in that time and place. If Jesus had been born in Hawaii rather than Bethlehem, no one seeing him standing up on the water would have jumped to the conclusion that he was a ghost. They would have assumed he was just another stand-up paddler.

27

 Curing a Leper without Antibiotics

After leaving Capernaum, Jesus goes on a preaching and demon-tossing tour, but things take a turn that presses the Messiah into practice as a healer.

> A leper came to him begging him, and kneeling he said to him, "If you choose, you can make me clean." Moved with pity, Jesus stretched out his hand and touched him, and said to him, "I do choose. Be made clean!" Immediately the leprosy left him, and he was made clean. After sternly warning him he sent him away at once, saying to him, "See that you say nothing to anyone; but go, show yourself to the priest, and offer for your cleansing what Moses commanded, as a testimony to them." (Mark 1:40–44, NRSV)

This episode is at once fascinating and odd. Very odd. Leprosy is an infectious disease caused by slow-moving mycobacteria. These rod-shaped germs take years or even decades to cause symptoms. When they finally get around to producing damage, it is primarily to the skin, mucosal linings, and eyes. What's odd about Jesus curing a leper?

For starters, the chances are pretty good that the supplicant was not in fact a leper. The dopey mycobacteria aren't especially good at leaping from one victim to another. Plenty of other, more common skin conditions, from acne or chickenpox to warts or eczema, might have been classed as leprosy. There were no med-school grads running around the Holy Land performing diagnoses. Getting classed as a leper was tough luck, but it didn't necessarily mean you were sick, or if you were you might well have a self-curing condition like poison ivy.

Writing in the Jewish Encyclopedia, Emil Hirsch and colleagues state:

> There is much reason to believe that the segregation of lepers was regarded, at any rate at certain periods, more in the light of a religious ceremonial than as a hygienic restriction. Zara'at was looked upon as a disease inflicted by God upon those who transgressed His laws, a divine visitation for evil thoughts and evil deeds. Every leper mentioned in the Old Testament was afflicted because of some transgression

Rather than doctors and diagnoses, there were Jewish priests and Jewish law with its obsessive concerns for *ritual* cleanliness. Against this backdrop, both authentic cases of leprosy and a wide variety of skin ailments and mere discolorations may have rendered unfortunate people outcasts. Once a person was declared taboo, the only way back was through a Jewish priest's ritual cleansing. That's why, after performing a "cleaning," Jesus is quoted as telling the "leper" to go to a priest. Without that additional step, the leper would be like a guy who's finished law school but hasn't taken the bar.

But let's back up. What's special about this story? For one thing, it's the earliest recorded instance of Jesus as a healer. Curiously, though, it's not the first instance of a leper being healed. Again, if this is meant to mark Jesus as the Christ, why is it not unique? Come to that, why is a mere prophet able to cure Naaman of his leprosy *without even meeting the man*? The client is a general, for Pete's sake, and Elisha practically phones it in.

> Naaman came with his horses and chariots, and halted at the entrance of Elisha's house. Elisha sent a messenger to him, saying, "Go, wash in the Jordan seven times, and your flesh shall be restored and you shall be clean." (2 Kings 9–10, NRSV)

You gotta pity anyone doing their laundry downstream of the general that day. But does it make any sense that Elisha performs the deed more airily than Jesus? Perhaps. The contrast can be read as a sign that Jesus values compassion over the law. To touch a leper was forbidden, but that's what Jesus does. (Although we now know the risk of infection through casual contact with a leper to be low, no one then would have had the advantage of such knowledge. A taboo served to try to stop epidemics, even if it was often misapplied.) Anyway, a liberal, figurative reading of the text can indeed make sense of this passage as demonstrating the compassion of Jesus.

Our charge, however, is to read the text as describing a miracle and then try to find a naturalistic explanation that fits. With that in mind, how could a touch heal? It could only do so if the man suffered from social rather than bacterial leprosy. People instinctively fear and reject strangers who appear to be infected. Research in evolutionary psychology finds that people are more willing to "reciprocate trust from healthy-looking social partners than from social partners who are relatively unhealthy-looking." To be branded a leper would make one a social outcast, "like walking around in an oversized T-shirt with a Microsoft Windows print," to quote Aaron Sleazy's online "Guide to Dressing Like a Loser."

Yeah, bro'—especially if it's a Windows Vista T-shirt.

Of course, if a man really suffered from an incurable and readily transmissible skin disease, avoiding him would be a smart move. However, with an ill-defined skin disease in an ancient society in which bathing was a religious ritual rather than a daily habit, the risk of being mistaken for a leper was high. If you've ever seen an unfortunate, unwashed homeless person sitting on a city corner, ask yourself this: could you tell at a glance whether the person had a skin disease or was just filthy? Yet, the disheveled homeless person was Joe-average in the Holy Land.

Remember, too, that the concept of illness in biblical times was entirely different. The Israelites knew nothing of bacteria, viruses, or natural causes of disease. To them, both the cause and the cure were religious.

So, surely the touch of a man regarded by the mob as a blessed healer—and the Messiah by some—would be enough to resolve the question of disease versus dirt. A mere touch from Jesus would convince the crowd. Assuming the poor slob did not actually have a mycobacterial infection, then, a verdict in favor of filth would be a blessing in disguise.

And speaking of disguise, why did Jesus tell the man to keep his trap shut? The "compassionate act" reading of this episode would suggest modesty. Jesus was trying to be kind, yet didn't want to get attention for it. But miracles are *supposed* to be signs.

What's the point of a sign that nobody sees?

Fortunately for the Gospels, this guy races around Galilee, yelling, "Yippee! I'm now leprosy-free, thanks be to Jesus"—or its Aramaic equivalent. Fame has its price, and soon the nobody from Nazareth is mobbed by people begging for cures. According to the synoptic gospels—Mark, Matthew, and Luke, the three that largely overlap—Jesus then goes on a healing tear.

I don't want to bore you, dear reader, by reviewing each and every miracle cure, so gird your loins and I'll pass over the others. Instead, let's consider the most curious feature of all—the need for any miracle cures at all. Jesus, according to standard Christian theology, is the son of God, on a mission from that all-powerful, all-knowing, and supremely virtuous being. So, rather than waste time healing a leper here and ten lepers there (viz. Luke 11), wouldn't it have been better to hand out antibiotics? Especially to those who were infected but presymptomatic? Why wait twenty centuries for some preventive care?

Today, we know that timely intervention with appropriate medicine can cure leprosy (and many another disease) before it inflicts permanent damage. You'd think an all-knowing, all-powerful God would have some on hand.

28

 An Oversupply of Demons and a Handy Herd of Swine

For Jesus, casting out demons becomes practically a daily routine, but in this he was far from alone. Due to an abundant supply of malignant spirits in ancient Israel, there was plenty of work for those in the demon-removal business. Exorcism was the pest control of its time, but instead of spraying chemicals practitioners cast spells. Rabbi David Wolpe tells us:

> Jewish demons, like their counterparts in other traditions, like to inhabit people or simply upend them from time to time. Not only are there many discussions of demons in rabbinic literature, but also, as a result of demonic activity, there are many spells directed against them, as where there are demons, there must be defenses and antidotes.

Not even Jesus could meet all the demand, so he subcontracted some of the business to his apostles. According to the earliest gospel, their job descriptions ran thus: "to be with him, and to be sent out to proclaim the message, and to have authority to cast out demons." (Mark 3:14–15)

So what makes exorcism by Jesus special? That's easy: style.

You know the old disappearing coin trick, right? Anybody with at least three fingers and a palm can pull that off. Your average professional magician tops that by making a person vanish from a cabinet. But to set himself apart, illusionist David Copperfield gathered an audience on Liberty Island, and before a live TV audience made the Statue of Liberty disappear. (See it on YouTube here: https://youtu.be/823GNH4Rczg.)

In much the same way, when Joe ben Schmoe casts out a demon, it's one and done. But in the Gadarene Swine affair, Jesus shows how to exorcise *en masse* and with flair.

The Lord has been touring the Sea of Galilee, dishing out parables. After stilling a storm, he crosses the sea and comes, the Bible says, to the country of the "Gerasenes" or "Gadarenes." Either way, it's a puzzle for scholars, for there was no such hamlet anywhere near the Galilee. Since "Gadarene" has become a byword for a headlong rush into disaster, we'll stick with that.

> They came to the other side of the sea, to the country of the Gadarenes. And when he had stepped out of the boat, immediately a man out of the tombs with an unclean spirit met him. He lived among the tombs; and no one could restrain him any more, even with a chain; for he had often been restrained with shackles and chains, but the chains he wrenched apart, and the shackles he broke in pieces; and no one had the strength to subdue him. Night and day among the tombs and on the mountains he was always howling and bruising himself with stones. (Mark 5:1–5, NRSV)

It's interesting to note here that ancient Israel was millennia ahead of its time in deinstitutionalizing its mentally ill. It took the United States until 1995 to adopt policies leading to community-based accommodations for the psychotic, in places ranging from sidewalks and public parks to the White House. But back to our story.

> But when he saw Jesus afar off, he ran and worshipped him, And cried with a loud voice, and said, What have I to do with thee, Jesus, thou Son of the most high God? I adjure thee by God, that thou torment me not. For [Jesus] said unto him, Come out of the man, thou unclean spirit. And he asked him, What is thy name? And he answered, saying, My name is Legion: for we are many. (Mark 5:6–9, KJV)

Let's pause to note that this line, introducing a collective self, has itself become legendary, inspiring countless reiterations in literature, film, television, and, more recently, video games. There's a Wikipedia page dedicated to its many allusions. But back to the Bible. Here, a spokesspirit for the gang enters a plea to stay put:

> He begged him earnestly not to send them out of the country. Now there on the hillside a great herd of swine was feeding; and the unclean spirits begged him, "Send us into the swine; let us enter them." So he gave them permission. And the unclean spirits came out and entered the swine; and the herd, numbering about two thousand, rushed down the steep bank into the sea, and were drowned in the sea. (Mark 5:1–13, NRSV)

A most curious tale. It echoes through Matthew and Luke, but in this earliest version, in Mark, it finds its fullest expression. This poor guy is nuts. That much is obvious, but how does his condition connect with the pigs? What the ancients called demons were surely, in some if not all cases, microbes. Among these are many candidates for infectious mental disease, ranging from fast-acting rabies to slow-burning syphilis. Recent research suggests that schizophrenia, whose symptoms closely resemble the story's "legion" of demons, may also develop through viral infection.

What's more, numerous infectious diseases, known as zoonotic, can leap between animals and humans. Better still, among zoonotic host animals, none figures more prominently than the pig. So, we may be able to reseat the story in the context of mental illness, animal husbandry, and zoonotic diseases. Yet, this hardly amounts to a satisfactory explanation for the tale. Actually, no one can make literal sense of it. Scholars are stumped not only about the oddity of the dual place names but also the curious fact that while Mark presents one madman, in Matthew there are two. Digging a little deeper, Dr. Russell Morton of the Lutheran School of Theology at Chicago notes that the writer doesn't seem to know what he's talking about:

> Pigs are not strictly herbivores but omnivores, and while they do eat various plant material, such as roots, berries, etc., their behavior is not normally described as "grazing." It is, rather, "rooting." This behavior was noted in the ancient world.

Even if you cut the Bible a break on this, Morton says, the problems continue:

> While the author of Mark may have described the feeding habits of the pigs with imprecise, popular language, the description of the stampede of the pigs is not so easily explained. . . . Pigs do not occur in large herds; in fact, the term twice used in Mark (5:11,13) to describe pigs is always used of oxen or [cattle] in Homer.

So there you have it. Swine root rather than graze. They don't form large herds, and they're not known for stampeding. But the problems don't stop there. This is ancient Israel. Jesus is preaching to observant Jews. According to God's law, pigs are unclean. So what's a huge herd of swine doing there?

The best explanation for the story of the Gadarene Swine lies outside our writ. It's an allegory. What kind of allegory? Well, that depends on whom you ask. For a sappy Sunday sermon, we turn to Rev. Dr. Graham Dodds of the Wells Cathedral:

> This man has been completely defined by what assails him, by what robs him of joy and health, by what hinders him and keeps him bound, by all those things that keep him from experiencing life in its abundance. And here's the thing: I think a lot of us aren't all that different.

Sure, apart from being a raving loony and living in a graveyard, the Gadarene Swine guy is just like the rest of us. Dodds quotes a colleague to elaborate the point:

> We talk about confronting our demons, but what do we mean by that? I think we mean confronting the things that dehumanise us and prevent us from being who we really are. Things that worry us, things that make us fearful—things that destroy our self-worth and self-esteem.

Really? Our everyday worries make us like a chain-breaking, graveyard lunatic who raves, howls at the moon, and beats himself with stones? C'mon, fellas, you're making the biblical literalists look right smart.

A much more sensible and tight-fitting explanation comes from Christian authors Shane Clairborne and Chris Haw: the story of the Gadarene swine is a satirical allegory about the hated Roman occupation.

> Jesus asked the man his name, and he replied, "Legion," the same word for a division of Roman soldiers. Scholars note that a legion consisted of around two thousand troops, and there would have been several legions around the Decapolis. It's interesting that in the story, the demons beg to stay in the area. Nearby was a "band" of pigs, band being the same word used for a group of military cadets. . . . The demons asked to be sent among the pigs, another symbol of uncleanliness. (Jews did not touch pigs.) Jesus invited the Legion to enter the pigs. And the pigs, specifically numbered at two thousand, "charged" into the sea to their deaths. And none of the listeners could have missed the subversive poetry, remembering the legion of Pharaoh's army that charged into the sea, where they were swallowed up and drowned (Exodus 14).

Our last resort, then, is to reach for a psychological explanation for the Gadarene Swine: the ability of the human mind to take an existing story about a stampede of cattle, transform and encode it, and let it reemerge as a tale that would surely would be read as an allegory about Jesus the Messiah, overthrowing piggish Roman rule.

Why . . . set it to music, and it could be a Negro spiritual about the hated slave masters, suitably disguised as hogs.

29

 A Fishy Tale

Fish swim throughout the New Testament and spill into Christology beyond. As anyone who's ever seen a car bumper knows, the fish is a symbol for Christ. What is perhaps not so well known is the reason. It's an acronym, or more specifically, a *backronym*—an existing word or words whose letters are subsequently used to form a phrase. An extravagant instance is the USA PATRIOT Act, passed in response to the terrorist attacks of September 11, 2001. In rushing through a 342-page bill that most members had not read, Congress strained mightily to make the USA PATRIOT Act stand for "Uniting and Strengthening America by Providing Appropriate Tools Required to Intercept and Obstruct Terrorism."

So how did Jesus come to be associated with a fish? Through the Greek word for it, *ichthys*. Using Greek letters, this provides the backronym *Iēsous Christos, Theou Yios, Sōtēr*, which translates into English as "Jesus Christ, Son of God, Savior."

Now, you might think that this being so, Jesus would show a little reverence toward fish. But no. Apart from telling his first disciples that they were to be "fishers of men," Jesus shows a decidedly indifferent attitude toward our finny neighbors in the sea. At most, they serve as props for his parables and miracles.

In Matthew, for example, Jesus offers a "good fish, bad fish" parable:

> Again, the kingdom of heaven is like a net that was thrown into the sea
> and caught fish of every kind; when it was full, they drew it ashore, sat

down, and put the good into baskets but threw out the bad. So it will be at the end of the age. The angels will come out and separate the evil from the righteous and throw them into the furnace of fire, where there will be weeping and gnashing of teeth. (Matthew 13:48–50, NRSV)

Not a pleasant fate if you're a fish—or a sinner. A few chapters later, Jesus again displays a codswalloping attitude. You recall I mentioned the old trick of producing a coin from someone's ear? Well, fish don't have ears, but that doesn't mean you can hurt their feelings.

When they reached Capernaum, the collectors of the temple tax came to Peter and said, "Does your teacher not pay the temple tax?" He said, "Yes, he does." And when he came home, Jesus spoke of it first, asking, "What do you think, Simon? From whom do kings of the earth take toll or tribute? From their children or from others?" When Peter said, "From others," Jesus said to him, "Then the children are free. However, so that we do not give offense to them, go to the sea and cast a hook; take the first fish that comes up; and when you open its mouth, you will find a coin; take that and give it to them for you and me." (Matthew 17:24–27, NRSV)

Pretty slick. And yet, we have to take it on faith that a miracle followed. Once Jesus dispatches Simon to win the fish lottery, the gospel moves on. But suppose things happened that way: would it require a miracle? Strictly speaking, it would not. That a fish would swallow a coin produces no gasps; to this day we use shiny lures to catch fish. To catch a fish that has swallowed a coin therefore requires no more than good luck. If, for example, there were 100,000 large fish in Capernaum's waters, and if twenty coins a year were dropped in those waters and half were swallowed by fish, the odds of retrieving a coin that way might be as low as 1 in 10,000. Now, those are long odds, but hey, people win lotteries against vastly longer odds (commonly, 1 in 183 million). Still, it would be quite surprising if someone could accurately predict catching a fish with a coin in its gullet.

That's where the possibility of postdiction comes in. To be sure, this would be a major deviation from the Bible, but we have to acknowledge that the event—finding a coin in a fish's mouth—could have happened, and then in trying to decide its meaning the cause might well have been attributed back to Jesus. It's human nature to give intentional cause preference over mere chance. We are agency-seeking creatures.

From there, the weaving of a little narrative to explain the event would be the most natural thing in the world. We are also storytelling creatures. If this were how things came about, Simon finding a coin in the mouth of a fish would be only a mild surprise. Then again, it's surprising when someone pulls a coin from your ear—at least the first time.

Fish keep turning up in the New Testament. When a football team wins a championship, they go to Disneyland. When the Apostles see their Lord crucified and resurrected, they go fishing. And guess who shows up to make the day?

> Simon Peter said to them, "I am going fishing." They said to him, "We will go with you." They went out and got into the boat, but that night they caught nothing. Just after daybreak, Jesus stood on the beach; but the disciples did not know that it was Jesus. Jesus said to them, "Children, you have no fish, have you?" They answered him, "No." He said to them, "Cast the net to the right side of the boat, and you will find some." So they cast it, and now they were not able to haul it in because there were so many fish. That disciple whom Jesus loved said to Peter, "It is the Lord!" When Simon Peter heard that it was the Lord, he put on some clothes, for he was naked, and jumped into the sea. But the other disciples came in the boat, dragging the net full of fish, for they were not far from the land, only about a hundred yards off. (John 21:3–7, NRSV)

That wacky Simon Peter! Why was he fishing in the nude to begin with? We can only speculate—and hope that there were no spiny fish in the sea. Once they've hauled in their net, the rest of the disciples follow, presumably in full dress. The Lord has his own catch, and his blessed fish fry is already under way.

> When they had gone ashore, they saw a charcoal fire there, with fish on

it, and bread. Jesus said to them, "Bring some of the fish that you have just caught." So Simon Peter went aboard and hauled the net ashore, full of large fish, a hundred fifty-three of them; and though there were so many, the net was not torn. Jesus said to them, "Come and have breakfast." Now none of the disciples dared to ask him, "Who are you?" because they knew it was the Lord. Jesus came and took the bread and gave it to them, and did the same with the fish. This was now the third time that Jesus appeared to the disciples after he was raised from the dead. (John 21:9–14, NRSV)

Any way you look at it, John 21 is a weird chapter. For one thing, this is not the third, but the fourth time Jesus appears. Go figure.

But, mathematical and sartorial puzzles aside, why would all the fish remain on the right throughout the night? For that matter, why would the disciples fish only on the left side of the boat? Why can no one recognize Jesus? But most of all, why would the Son of God, just back from the dead, spend a night playing a prank on his key followers?

The only way to make sense of this is to read it as a parable. Certainly, it is a passage rich with symbolism: fishing in the darkness produces no catch, but when the light dawns . . . The left side is sinister, but the right side is, well, right. It's also a reminder that Jesus first gathered his disciples from among fishermen, and that on various occasions in the Gospels he provides abundance, so remember, wayward Christians, don't bite the hand that feeds you.

But fishing for symbolic meanings is not our task. Leaving exegesis to others, let's consider these events as possibly natural occurrences. Do fish congregate in one spot rather than another? Sure. Most fish swim in schools, and together they find their feeding grounds. Could Jesus, standing on the beach, have spotted a school of fish on the opposite side of the boat? Mmmm, tricky, that one—but not impossible. If all the disciples were on one side of the boat, handling the nets, and all the fish were on the other, Jesus might have seen what the others could not: that when fish are close to shore they frequently jump out of the water momentarily to throw off predators.

Could Jesus have donned a disguise (maybe an older man?) to pull this prank? Yes, indeed, and that might explain why he calls them "children." As for figuring out who was advising them, the disciples might not have been able to see through the disguise, but they quite likely could have recognized his voice. Finally, could Jesus have risen from the dead? Well, that's for another chapter, and before that we have other fish to fry.

30

 The Meanest Miracle

If you think Jesus was unkind to the coin-swallowing fish, wait till you see how he treats a poor, innocent fruit tree.

> In the morning, when he returned to the city, he was hungry. And seeing a fig tree by the side of the road, he went to it and found nothing at all on it but leaves. Then he said to it, "May no fruit ever come from you again!" And the fig tree withered at once. When the disciples saw it, they were amazed, saying, "How did the fig tree wither at once?" Jesus answered them, "Truly I tell you, if you have faith and do not doubt, not only will you do what has been done to the fig tree, but even if you say to this mountain, 'Be lifted up and thrown into the sea,' it will be done. Whatever you ask for in prayer with faith, you will receive." (Matthew 21:18–22, NRSV)

You're probably familiar with the conventional lesson that Sunday School teachers draw from this: faith moves mountains. But there are several peculiar features here. First, let's be honest: this amounts to a messianic hissy fit. I guess we should be grateful that Jesus didn't stop by a McDonald's and learn that they were out of Egg McMuffins.

But consider the fruit of the lesson: either nobody has doubt-free faith, or Jesus was just plain wrong—to date, no one has demonstrated the ability to toss a mountain into the sea via prayer. The act of prayer may have salutary effects on the person praying—much as meditation does. But attempts to alter the course of events via prayer show no consistent evidence of success.

One general indicator: before the invention of antibiotics and vaccines but during an era when prayer was more common, one in seven Americans died of tuberculosis. Today, despite the decline of prayer, almost no one dies of tuberculosis in the United States. (The death rate has dropped from 1 in 7 to about 1 in 600,000, according to CDC figures.)

The most systematic attempts to study the efficacy of prayer have focused on third-party prayers for the health of someone else. These so-called intercessory prayers are hard to study, because researchers can never be sure who is praying for whom, and what is being asked. Members of the control group (for whom no prayers are slated) might be prayed for by well-wishers outside the study. Those in the experimental group (toward whom prayers are directed) might be the targets of malicious prayers. Worse yet, the experimental design has often been flawed, and in at least one instance, fraudulent.

Nevertheless, numerous studies have been attempted, with mixed results. The most rigorous, well-designed, and carefully done study, funded by the pro-religion John Templeton Foundation, yielded an alarming result: not only did prayer have no positive effect on recovery from heart surgery, but also those patients who were informed that they would be the target of prayers suffered more complications.

In short, the studies fail to confirm the efficacy of prayer. As for the extravagant claim that Jesus makes, Hector Avalos, a professor of religion who was raised as a child preacher and later turned skeptical, has a thing or two to say on that point. He points out that when an airplane fails in flight the longer a passenger has time to pray the lower his or her odds of survival become. Low-altitude crashes, produced by, say, engine failure on takeoff, often produce survivors. High-altitude crashes almost never do. This leads Professor Avalos to wonder aloud: does God hate high-flyers?

Moreover, if the point of blighting a tree was to teach about prayer, why such a petulant example? Why not make the tree bust out in figs, for example? You know, keep it positive.

No one else has shown an ability to wither a tree—at least not until Agent Orange came along during the Vietnam War. But if the normally compassionate Jesus were packing any kind of defoliant you'd expect it to be something a little less lethal. Maybe a nice weed suppressor like Roundup. After all, if he knew in advance that the fig tree wouldn't have any fruit on offer, he needn't have bothered to visit. It's a puzzler, all right.

Finally, why would God require belief as a prerequisite for granting favors? Belief itself is not a moral good. Some people sincerely believe that God wants them to tape explosives to their bodies and detonate them in public places. That's not good. . . . Right?

Other people believe in helping the poor or saving the whales. In short, people believe all sorts of things, good, bad, and downright weird. Not that whales are weird. Some people believe that getting a Justin Bieber tattoo is a decision they won't bitterly regret.

"Hold on," saith the preacher. "This isn't about some Canadian pop star. Read your Bible! That's what God wants us to believe." Okay, but what about flat-Earth believers? They rely on biblical authority for their counterfactual belief. Does that mean they should go to the head of the prayer-line?

Some might say God just wants us to believe in *him*—that's where faith comes into it, and with faith, moral goodness. If that's the case, then Moses was not especially good, and Joshua was downright rotten. For them, God makes it easy: he pops up all over the place. Burning bush here, temple dweller there. He led the entire Hebrew nation into the wilderness from atop a cloud. In short, he *revealed* himself. If God wants people to believe in him, he doesn't need to play peek-a-boo. He just has to show himself in a consistent and unmistakable way.

And leave the fig trees be.

31

 You Can't Keep a Good Man Down

Unless you've lately arrived from Mars, or maybe Pyongyang, God's plan for the Big Finish will come as no surprise: Jesus must be sacrificed on the cross to redeem our sins. The most famous statement of this, the one that's always held up in stadiums by evangelical sports fans, comes in John 3:16: "For God so loved the world, that he gave his only Son, that whoever believes in him should not perish but have eternal life."

But why? That's a bit of a head-scratcher. The Old Testament is chock full of sacrifices to the Lord—but all of them involve choice animals, ritually slaughtered, roasted, and (we suspect) later consumed by priests. True, God orders Abraham to sacrifice his son Isaac, but turns out he's not serious. It's just a gruesome loyalty test.

> When they came to the place that God had shown him, Abraham built an altar there and laid the wood in order. He bound his son Isaac, and laid him on the altar, on top of the wood. Then Abraham reached out his hand and took the knife to slaughter his son. But the angel of the Lord called to him from heaven, and said, "Abraham, Abraham!" And he said, "Here I am." He said, "Do not lay your hand on the boy or do anything to him; for now I know that you fear God, since you have not withheld your son, your only son, from me." And Abraham looked up and saw a ram, caught in a thicket by its horns. Abraham went and took the ram and offered it up as a burnt offering instead of his son. (Genesis 22:9–13, NRSV)

Ah, that Yahweh—what a kidder. But note that if God had meant it, the sacrifice was to be performed in a sacred manner, at an altar, with a ritual throat slitting, followed by burning. Compare that with what happens to Jesus:

> Then the soldiers . . . stripped him and put a scarlet robe on him, and after twisting some thorns into a crown, they put it on his head. They put a reed in his right hand and knelt before him and mocked him, saying, "Hail, King of the Jews!" They spat on him, and took the reed and struck him on the head. After mocking him, they stripped him of the robe and put his own clothes on him. Then they led him away to crucify him. (Matthew 27:27–31, NRSV)

Notice any difference from your standard holy sacrifice? Far from an ennobling ritual, this is debasement and humiliation at its worst. Though a resurrection scene is to follow, we still have to ask: does this make any sense as a divine plan? Why, for instance, crucifixion?

Today, the cross serves as the chief symbol of Christianity, but at the time it was the go-to punishment for high crimes and (political) misdemeanors. It was the Romans' way of letting a person know they were the lowest of the low. After the slave rebellion led by Spartacus, for example, some 6,000 slaves were put to death by crucifixion. It was a means of death designed to be slow, humiliating, and extremely painful. So, in crucifying Jesus, the Roman authorities were not setting Jesus apart for special treatment, but indicating that he was, in the words of another empire (the one George Lucas imagined), "rebel scum."

Remember, one thing all Christians agree on is that Jesus was human. He may or may not have also been divine, but his human side was fully capable of suffering. The Gospels make this explicit:

> [A]bout three o'clock Jesus cried with a loud voice . . . "My God, my God, why have you forsaken me?" . . . Then Jesus cried again with a loud voice and breathed his last. (Matthew 27:46–50, NRSV)

Game over. His defeat, humiliation, and indeed his life, are complete. Or so it seems. But then comes the miracle. It's one we'll have to try to explain, but first we have to come to grips with the story.

The gospel accounts of Jesus' resurrection differ in many ways, but in broad outline they agree. Three of the four have Jesus repeatedly foretell his betrayal, death, and resurrection—which makes the death scene quoted above even more puzzling. If Jesus was in on the plan, where does "forsaken" come in?

Back to the miracle. Matthew sets it up with a bit of "nothing up my sleeves" business:

> When it was evening, there came a rich man from Arimathea, named Joseph, who was also a disciple of Jesus. He went to Pilate and asked for the body of Jesus; then Pilate ordered it to be given to him. So Joseph took the body and wrapped it in a clean linen cloth and laid it in his own new tomb, which he had hewn in the rock. He then rolled a great stone to the door of the tomb and went away. Mary Magdalene and the other Mary were there, sitting opposite the tomb. (Matthew 27:57–61, NRSV)

So, we have witnesses in place to assure that there's no funny business going on at the tomb, no body snatching. But you know how the Bible feels about women: they're lucky even to have names. If there's to be any controversy, we'll need better witnesses than these two. Time to bring in security professionals:

> The next day, that is, after the day of Preparation, the chief priests and the Pharisees gathered before Pilate and said, "Sir, we remember what that impostor said while he was still alive, 'After three days I will rise again.' Therefore command the tomb to be made secure until the third day; otherwise his disciples may go and steal him away, and tell the people, 'He has been raised from the dead,' and the last deception would be worse than the first." Pilate said to them, "You have a guard of soldiers; go, make it as secure as you can." So they went with the guard and made the tomb secure by sealing the stone. (Matthew 27:62–66, NRSV)

This scene strikes many scholars and knowledgeable people as highly improbable. As Reza Aslan points out in his book *Zealot*, once the Romans crucified someone, especially a political upstart, he stayed put. "Because the entire point of crucifixion was to humiliate the victim and frighten the witnesses, the corpse would be left where it hung to be eaten by dogs and picked clean by birds of prey." Oy.

As for Pilate taking pity on Jesus in particular, this is equally improbable. He had routinely slaughtered Jewish "troublemakers," and would go on doing so for the rest of his term as prefect (or "procurator") of Judaea, which still had several years to run. Crucifixions were part of his daily routine. "If Jesus did in fact appear before Pilate, it would have been brief and, for Pilate, utterly forgettable," writes Aslan. Even granting for the moment that Pilate had planned to release someone for Passover, what are we to make of this gospel account?

> Pilate replied, "I am not a Jew, am I? Your own nation and the chief priests have handed you over to me. What have you done?" Jesus answered, "My kingdom is not from this world. If my kingdom were from this world, my followers would be fighting to keep me from being handed over to the Jews. But as it is, my kingdom is not from here." Pilate asked him, "So you are a king?" Jesus answered, "You say that I am a king." (John 18:28–40, NRSV)

For someone representing himself, Jesus proves an adept defense lawyer: by denying that he's an earthly king, Jesus avoids guilt of sedition under Roman law. Blaspheming against Jewish law is no crime as far as Pilate is concerned. He declares him innocent.

But the procurator's subsequent actions are baffling. If Jesus is innocent of violating Roman law, Pilate should either free him or return him to the Jewish authorities to carry out their own trial. As for granting a pardon for Passover, by definition that had to be awarded to a convict. You can't *pardon* the innocent; you can only release them.

In any event, the notion that Pilate would have told a Jewish mob, "Here, you decide" seems out of character. The most infamous version appears in Matthew:

> So when Pilate saw that he could do nothing, but rather that a riot was beginning, he took some water and washed his hands before the crowd, saying, "I am innocent of this man's blood; see to it yourselves." Then the people as a whole answered, "His blood be on us and on our

children!" So he released Barabbas for them; and after flogging Jesus, he handed him over to be crucified. (Matthew 27:24–26, NRSV)

Yeah, right. The raucous crowd demanded to be cursed down the generations, and maybe cried out for higher taxes, fewer holidays, and more potholes in the roads. The "blood curse" indeed cropped up. In centuries to come, as Christians became distinct from Jews and eventually took over the Roman Empire, this passage would be invoked to justify an unrelenting persecution of the Jews. But the historical Pilate was ahead of the curve. He had no qualms persecuting Jews.

The actual procurator was notable for treading on Jewish religious toes, and his reaction to riots was never known to display the light touch. Two Jewish historians of that era, Philo and Josephus, describe Pilate as anything but hesitant or thoughtful in character, and wholly contemptuous of the Jewish religion. Philo, according to the Jewish Encyclopedia, wrote of Pilate that "his administration was characterized by corruption, violence, robberies, ill treatment of the people, and continuous executions without even the form of a trial." Josephus reports that when Pilate arrived in Judaea, he put Roman symbols up all over Jerusalem. Annoyed by frequent Jewish protests, the prefect devised a novel strategy for crowd control. Josephus reports that, after Pilate helped himself to funds from the Temple treasury for improvements to the aqueduct, yet another demonstration broke out.

Now when he was apprized aforehand of this disturbance, he mixed his own soldiers in their armor with the multitude, and ordered them to conceal themselves under the habits of private men, and not indeed to use their swords, but with their staves to beat those that made the clamor. . . . Now the Jews were so sadly beaten, that many of them perished by the stripes they received, and many of them perished as trodden to death by themselves; by which means the multitude was astonished at the calamity of those that were slain, and held their peace.

Eventually, in 36 CE, Pilate's bloody suppression of the local population led to his recall to Rome. Somewhere along the line, though, the Bible would have us believe, the Jews not only assembled but also started to riot, and that Pilate meekly responded by giving them Barabbas.

Well, let's say that's so. Barabbas lived to rob another day, while Jesus got nailed to a cross. Then what?

Even if Jesus had been lucky enough to have a decent burial, the Roman authorities would have had no worries. Pilate would have slept like a baby and kept his guards close by in case he needed to beat up some more Jewish supplicants. As for troublesome upstarts, religious zealots were a denarius a dozen back then. Nailing them one by one to a cross had proven an effective means of rule—and would continue to be so for centuries to come.

But, like Pilate's improbable handwashing, the guards have a role to play in this narrative: they serve to quell any doubts about the miraculous nature of the scene that is to come.

32

 Hey! Where'd He Go?

And so, a really bad week for Jesus comes to an end. But things are about to change. Turns out God has scripted a Disney Death. I'm not making this up; that's a genuine term of art in the screenwriting business. Following a heroic sacrifice the audience is sniffling and wiping away tears when, surprise! The hero revives! It's not a downer ending after all. Now, if this were a Disney movie, someone would have to kiss Jesus to bring him back to life. But that's not the Bible's way:

> After the sabbath, as the first day of the week was dawning, Mary Magdalene and the other Mary went to see the tomb. And suddenly there was a great earthquake; for an angel of the Lord, descending from heaven, came and rolled back the stone and sat on it. His appearance was like lightning, and his clothing white as snow. For fear of him the guards shook and became like dead men.
>
> But the angel said to the women, "Do not be afraid; I know that you are looking for Jesus who was crucified. He is not here; for he has been raised, as he said. Come, see the place where he lay. Then go quickly and tell his disciples, 'He has been raised from the dead, and indeed he is going ahead of you to Galilee; there you will see him.' This is my message for you." So they left the tomb quickly with fear and great joy, and ran to tell his disciples. Suddenly Jesus met them and said, "Greetings!" (Matthew 28: 1–9, NRSV)

Isn't it odd that this angel, who has just come flying out of the sky, flashing like lightning, has to strain and heave to roll away a great big boulder just to convince these ladies that Jesus is up and at 'em? Personally, I might have been inclined to take his word for it. But in any event, the Lord himself pops up out of the bushes a few minutes later, so it really seems like a wasted angelic effort.

This is one of several curious details in the resurrection narrative that suggest the writers themselves were straining to overcome the skepticism of their audience. I can hear the talk-back session at a public reading.

"Right. An angel *told* 'em?" yells someone in the crowd. "Ehh, g'wan wit' youse."

"No, but wait," rebuts the storyteller. "The angel didn't just tell them. He rolled away the boulder and *showed* them. They could see for themselves that the tomb was *empty*."

"Oh, yeah? I suppose the guards just let this happen and then marched off to report to Pilate."

"No, look. The guards were all like 'ubudda, ubudda' and their knees were clashing like cymbals. I'm telling you, they went into shock the moment they saw this angel swooping down on them like a UFO."

"Eh? What's a UFO?"

"Errrm, uh, you know, an unidentified flapping object."

Is Matthew 28 a rebuttal? Well, consider that in Mark, the earliest written version, the stone has already been rolled away by the time the ladies show up. What's more, the text makes it sound like the gardener could roll away the stone. Don't forget: Joseph of Arimathea, who was no spring chicken, rolled it there in the first place. Even weirder, accompanying Mary Magdalene and Mother Mary is a woman of distinctly ill repute:

> When the sabbath was over, Mary Magdalene, and Mary the mother of James, and Salome bought spices, so that they might go and anoint him. And very early on the first day of the week, when the sun had risen, they went to the tomb. They had been saying to one another, "Who will roll away the stone for us from the entrance to the tomb?" When they looked up, they saw that the stone, which was very large, had already been rolled back. As they entered the tomb, they saw a young man, dressed in a white robe, sitting on the right side; and they were alarmed. But he said to them, "Do not be alarmed; you are

looking for Jesus of Nazareth, who was crucified. He has been raised; he is not here. Look, there is the place they laid him. (Mark 16: 1–6, NRSV)

Salome? She of the seven veils? The scarlet lady who demanded and got the head of John the Baptist on a platter, with sprigs of parsley all around? Surely not. Scholars reckon this Salome is the wife of Zebedee, that poor old guy whose business Jesus ruined at the beginning of the Gospels by telling Z's sons to drop their fishing nets and follow him instead in the fishing-for-men trade. This Salome, the exegetes say, is their mom and a devoted follower of Jesus. Fair enough, but you can see why, to avoid confusion, she was edited out in later versions.

So far, what we have is (a) a missing corpse and (b) the reappearance of a walking, talking Jesus. But two of the gospels go on to stress that the returned Jesus is not a ghost.

In Luke, Jesus first appears to a couple of guys walking on the road to Emmaus, then shows up in Jerusalem at a gathering of his disciples:

Jesus himself stood among them and said to them, "Peace be with you." They were startled and terrified, and thought that they were seeing a ghost. He said to them, "Why are you frightened, and why do doubts arise in your hearts? Look at my hands and my feet; see that it is I myself. Touch me and see; for a ghost does not have flesh and bones as you see that I have." And when he had said this, he showed them his hands and his feet. (Luke 24:36–40, NRSV)

Presumably, this last gesture is meaningful because a ghost would not have puncture wounds left over from the crucifixion. Still, the disciples need more convincing, so Jesus comes up with proof positive—to be more specific, a piscine proof.

While in their joy they were disbelieving and still wondering, he said to them, "Have you anything here to eat?" They gave him a piece of broiled fish, and he took it and ate in their presence. (Luke 24:41–43, NRSV)

Meanwhile, over at John's gospel, the Lord has to convince an especially skeptical apostle who missed the first reappearance and suspects the guys are pulling a prank on him. Apparently, he doesn't buy the adage that seeing is believing:

> Thomas (who was called the Twin), one of the twelve, was not with them when Jesus came. So the other disciples told him, "We have seen the Lord." But he said to them, "Unless I see the mark of the nails in his hands, and put my finger in the mark of the nails and my hand in his side, I will not believe." (John 20:24–25, NRSV)

A week later, Jesus again returns and displays the winning touch:

> Then he said to Thomas, "Put your finger here and see my hands. Reach out your hand and put it in my side. Do not doubt but believe." Thomas answered him, "My Lord and my God!" Jesus said to him, "Have you believed because you have seen me? Blessed are those who have not seen and yet have come to believe." (John 20:27–29, NRSV)

Why is it so important to demonstrate that Jesus is not a ghost? For the gospel writers, one issue may have been that while seeing a ghost was an everyday phenomenon for their audience, they wanted to make it clear that what happened here was really exceptional. But there may be other reasons as well. According to Randy Alcorn, a prolific Christian author and founder of Eternal Perspective Ministries, the miracle would not be complete without the body:

> The physical resurrection of Jesus Christ is the cornerstone of redemption—both for mankind and for the earth. Indeed, without Christ's resurrection and what it means—an eternal future for fully restored human beings dwelling on a fully restored Earth—there is no Christianity. . . . A non-physical resurrection is like a sunless sunrise. There's no such thing. Resurrection means that we will have bodies. If we didn't have bodies, we wouldn't be resurrected!

If this makes theological sense, it's also convenient for science, as it rids us of the Casper-the-Ghost problem we noted back in chapter 3.

The conservation laws of physics give us every reason to believe that hallucination is the only explanation for perceiving a ghost. By definition, something immaterial cannot interact with photons or atoms to produce an image or sound. So, thank you, Jesus! With you in corporeal form, it'll be a whole lot easier to explain your resurrection.

33

 Death, What Is Thy Thing?

Even with a body on tap, we face one tough task. We have to establish that someone who was dead came back to life. What could that mean? First, we need to get a grip on what it means to die, and next what happens when we die.

By contemporary medical standards, death is the irreversible dissolution of the brain. A stopped heart may be a pretty good indicator of death, but in modern terms it's not enough. There are numerous documented cases of people recovering after 20 minutes or more without a heartbeat—provided they are kept very cool in the meantime. Hypothermia appears to protect the brain from devastation.

For that reason alone, heart stoppage is not a reliable criterion of death. But there is a deeper reason to adopt the standard of brain death. An enormous volume of research confirms that the "I" in each of us is generated within our brains. Functional magnetic resonance imaging—known as fMRI for short, reveals the brain in action. Researchers can see the differences between a brain concentrating on a math problem or listening to music. They can "see" lies taking shape in the brain. They can even tell what happens when a hipster gets high. And even if, as some say, half the fMRI studies will be thrown out in the next decade, one conclusion holds fast: hey, man, it's all organic.

The ancient notion of a soul separate from the brain finds no purchase on the mountain of evidence that we *are* our brains. Change the brain—with alcohol, antidepressants, or a stroke—and you change the self. Sever the corpus callosum—the bridge between the two hemispheres of the brain—and you effectively have two selves, although from the inside it may not feel that way. Only one half speaks, but the other can see, understand, and think. You can see a real-life demonstration of this on YouTube at https://youtu.be/ZMLzP1VCANo. The video introduces "Joe," a man who underwent split-brain surgery to alleviate terrible epileptic seizures he had suffered. Afterward, his mute right brain takes in information but cannot share it with his left brain, so when the right brain sees something, Joe can draw it but he cannot say the word for it. His speaking self is unaware that he's seen anything. So, does Joe now have two souls—one that speaks and one that remains silent?

Each of us has an intuitive, unshakeable sense of self. For the religious and spiritual among us, it feels like a soul, separate from the body. But all the evidence suggests there's no such thing.

We don't as yet know everything about the brain, and consciousness remains a deep problem, but we know with great confidence that electrochemical activity in the brain generates the self. It exists—there's no denying that—but all the evidence suggests that it emerges from within the brain, somewhat the way a movie emerges from a DVD when it is played.

Does the brain link up with a soul? It's an intuitively attractive concept, but one that holds no water. Consider the soul apart from the brain—a kind of immaterial, ghostly version of the self. What can it hear? Nothing. Sound is a pattern in the air that causes cilia in our ears to vibrate, sending signals that are interpreted by our brains as meaning, music, or noise. But a soul has no hearing apparatus.

What can it see? Nothing. Light is an electromagnetic wave racing through space. When certain frequencies strike our retinas, they cause signals that our brains interpret as sight. A soul has no retinas to slow the light waves down.

What can it do? Nothing. The laws of nature are such that energy is conserved. Immaterial things—if that concept makes any sense—cannot barge in and divert a single atom from its course. To do so would be to add extra energy to the universe and both theory and observation agree: that cannot happen.

You don't have to study physics to get this: if something pulls, pushes, or stops something else, it must involve energy or matter, which is really just a frozen form of energy. The immaterial affecting the material is a logical contradiction.

Can a soul at least remember what the body does? No. We now know that information processing—to record or retrieve a memory—takes energy, something the soul by definition does not have, 'cause it's immaterial.

Maybe we rescue the soul by redefining it. Suppose we think of it as a backup device, lodged in a parallel dimension, where it has energy of its own. There the soul sees, hears, feels, and remembers but remains completely passive. Maybe.

But, again, you don't have to be a physicist to doubt that the self is anything more than the brain. Consider what happens when you sleep. For roughly a fifth of each night's sleep, you completely lose consciousness. The self just isn't there. During REM sleep, so called for the rapid eye movements that accompany it, your brain generates dreams. In these, you are present, but it's a feeble version of yourself. The dreaming you cannot read, remember phone numbers, or reason its way through problems. But that's just as well, because your dream-self exists in an unstable world where no rules apply.

If you've ever undergone general anesthesia (not the twilight kind, where you remain half-awake), you probably know what death is like. In light of what we know about the self, it's most likely like nothing. Total lack of consciousness—no time passing, no dreams, no sensations, no bodily awareness. To a patient awakening after, say, a four-hour operation, it seems as if their eyelids shut one moment and reopened the next. In between . . . complete and utter nothingness.

But wait. What about all those stories of near-death experiences? The most extravagant of these include visits to heaven. Surely, that's evidence! Indeed it is: evidence that the brain, on the edge of anesthesia or hypoxia, can generate vivid and memorable hallucinations. We know that much is true. The "tunnel vision" that is a common feature of NDEs results from reduced oxygen flowing to the retina; the center has the greatest blood supply, therefore the periphery goes dark first. Likewise, the brightness that many report has a physical explanation: when short on oxygen the pupils relax and expand. As for floating out of the body or visiting heaven, to deny hallucination as an explanation is as unreasonable as to claim that when a person has a Peter Pan dream, they are actually flying. Remember, to float outside your body and see anything, you would have to intercept the light reflected off of objects. But with no physical you there, the photons would pass merrily on by. To hear anything you'd have to pick up vibrations in the air. No ears, no sound. Out of body, out of luck.

Back to Jesus. If we agree that when, and only when, his brain dies he is dead, is there any possibility of reversal of brain death? Medical experts would say no. After all, the very definition is "the *irreversible* loss of all functions of the brain, including the brainstem." Doctors now have pretty reliable tests for making that determination. Shining a light on the pupils is one. A failure to contract is a pretty good indication that the brainstem is out of commission. At the time of the crucifixion, of course, such tests did not exist.

Once death occurs, the body stiffens in what is called rigor mortis and then begins to decay. Cells that have been faithful servants to the self for a lifetime become our own worst enemies. Deprived of fuel, they shut down and their enzymes leak out and begin to corrode everything around them. The process is known as autolysis, which means, roughly, self-destruction. But if that weren't enough, bacteria, which are always present inside and on our bodies, go to town on what's left of us. About twelve hours after death, the combination of autolysis and putrefaction overwhelms rigor mortis.

As for the brain, once the neurons cease functioning the memories that make up a self are lost. Memory is another topic where science has a long way to go, but there is no doubt that it is physical. A gene we share with all mammals, Npas4, crucially guides the brain in preparing to store memories. When scientists delete the gene in lab mice, they are utterly unable to form memories. Some people suffer the same fate as a result of stroke. In a properly functioning brain, groups of neurons fire in a certain pattern, and subsequently re-create that pattern when the memory is recalled. But in brain death, by definition, memory is irretrievably lost. The information—and the intricate system that dynamically processes it—is dispersed into the environment as heat and other waste. That is why a corpse is so cold to the touch.

To come back to life three days after death would seem, then, to be impossible. Yet, there remains a possibility that Jesus could have made a magic-free comeback.

34

 All Stories Are True

A totally natural resurrection depends on the existence of something that is a matter of speculation. Yet, it is a speculation that many scientists now believe to be true: an infinite universe. There are several ways to conceive of an infinite universe. It might be an Everett multiverse, in which every quantum choice to be "here" or "there" results in the entire "universe" splitting in two. As absurd as that sounds, it is actually one of the least troubling explanations for what we observe of quantum behavior. No one can deny that in double-slit experiments, particles that are monitored by detectors in the slits seem to choose one path over another; while particles that go unobserved seem to ripple like waves through both slits. Can an electron or photon "choose"? Surely, it is easier to accept that *both* wave-like and particle-like movements happen, but we only get to see one result. This implies the kind of "many worlds" reality that Hugh Everett III proposed.

And that's not the only kind of multiverse that nature hints at. String theory, our best but as yet untested explanation for how general relativity and quantum mechanics fit together, implies a cosmic landscape of bubble universes, each with its own random set of initial conditions.

Whatever the architecture of the universe, if it is truly infinite, and if it allows for every possible configuration of energy and matter, and perhaps for every possible set of physical laws, then every outcome, no matter how wildly improbable, must take place. Yes, we have come full circle back to the totalitarian principle (mentioned in the introduction and again in chapter 1): anything that is not forbidden is mandatory.

Think about that. It means that all stories are true (all possible stories, that is). For starters, somewhere out there, far beyond the farthest galaxy we can see, beyond even what we know as time and space, is another bubble where Huck Finn and Jim really are floating down the Mississippi. Somewhere, there is a world on which a guy named Hitler started World War II, but then surprised everyone by giving an emotional broadcast speech in which he recalled his dear old Jewish grandmother, apologized for misleading his people, disbanded the SS, and convened world peace talks. If the laws of physics vary, somewhere out there is a bubble where Harry Potter really is Hogwart's most famous alum. (Maybe he vacations at Frodo Baggins World, a fantasy park in his world where "impossible" features of the *Lord of the Rings* trilogy are brought to life through the wonders of technology.)

As outlandish as these scenarios might seem, can we plausibly construct one in which Jesus comes back to life on the third day? Yes, we can. To do so, we must not only posit an infinite multiverse, we must also descend to the weird world of the quantum. Particles don't care about time. They go this way and that, jumping from one state to another, forward and backward in time, obeying only statistical laws.

In the macro world that we inhabit—the world of solids, liquids, plasmas, and gases—time makes its presence felt in a unidirectional way. As the poet W. B. Yeats wrote in his poem "The Second Coming":

Things fall apart; the centre cannot hold
Mere anarchy is loosed upon the world . . .

Writing in 1919, with the catastrophe of World War I and a failed Irish uprising ringing in his ears, Yeats was surely more concerned with politics than either religion or science. Yet, his words accurately describe the second law of thermodynamics: in a closed system, things tend toward disordered equilibrium. Entropy is the tendency of stuff to disperse. Left to itself, a teenager's bedroom drifts from low to high entropy over time. Just like the universe.

By definition, a universe is a closed system, so we expect that it will run down. And most systems within it will continually degrade. A typical physicist's example is an egg. If you drop one, there's no practical possibility of reversing course and putting it back into the shell. There are just too many ways for its molecules to be disordered, and only one way for them to line up and be an egg. That's why all the king's horses and all the king's men couldn't put Humpty together again.

To be sure, there are some whorls and eddies of order. Planet Earth, our solar-powered biosphere, constitutes Exhibit One. But the order that life and its extended phenotypes—termite mounds, eagle's nests, skyscrapers, airliners, and so on—exhibit come at a price of greater disorder overall. Ultimately, it appears, entropy wins.

Death is a local example. Although it doesn't normally come up in a freshman physics class, brain death is an excellent instance of a highly ordered structure becoming disordered and dispersing all of its stored information into the surrounding environment.

But here's the thing: entropy, too, is a statistical law. Flip a fair coin a million times, and it will come up heads exactly or very nearly 50 percent of the time. But somewhere in that million-flip sequence, you might well find a run of ten straight tails. So, what could conceivably solve the puzzle of resurrection? Chance.

If you roll a pair of dice long enough, you'll get double sixes. It shouldn't take you more than 24 tries.

If you shuffle a randomized deck of cards often enough, you'll be able to deal them in perfect order. That will likely take you far longer than the nearly 14 billion years our universe has been around. The odds of shuffling into perfect order are about 1 in 10^{67}.

And, if you root around in enough universe-bubbles in the multiverse, you'll eventually find one in which all the quantum particles that were the Lord's brain spontaneously reassemble three days after Jesus died. What are the odds against that? Staggering.

But in an infinite multiverse, somewhere there *must* be a bubble in which a "miracle" happens . . . just by chance. Actually, scientists have a scary thought experiment in which something quite similar happens. It's called the Boltzmann Brain, named after Ludwig Boltzmann, a nineteenth-century scientist who first formulated the entropy laws that drive any closed system downhill toward equilibrium. In a 2008 paper, physicists Andreas Albrecht and Lorenzo Sorbo explain:

A century ago Boltzmann considered a "cosmology" where the observed universe should be regarded as a rare fluctuation out of some equilibrium state. The prediction of this point of view, quite generically, is that we live in a universe which maximizes the total entropy of the system. . . . This means as much as possible of the system should be found in equilibrium as often as possible. From this point of view, it is very surprising that we find the universe around us in such a low entropy state. . . . The most likely fluctuation consistent with everything you know is simply your brain (complete with "memories" of the Hubble Deep fields, WMAP data, etc) fluctuating briefly out of chaos and then immediately equilibrating back into chaos again. This is sometimes called the "Boltzmann's Brain" paradox.

In other words, the simplest explanation for why you exist is that you are a clever bubble in the cauldron of chaotic reality—a bubble of consciousness and memory. But in an infinite universe even more improbable fluctuations must also happen, including the one we appear to be saddled with: a 5-billion-year-old world teeming with life. And somewhere out there must be outrageously improbable configurations of particles that happen to bring Jesus back to life.

And of course, if it's true for Jesus, it's true for everyone who ever lived. In an infinite multiverse, resurrection of the dead is guaranteed. With infinity, all possibilities must come to pass.

I don't know why anyone would find this idea appealing. To me, it is horrifying. If everything that can possibly happen necessarily happens, then meaning, choice, and fate are demolished. Every decision, choice, or chance is canceled out by its opposite. Stare into the abyss, my friend. Reality sums up to nothing.

It's a vision of the world that would make even a jaded French existentialist spit out his champagne. Fortunately, for now at least, it's only a speculation. When I had a chance to ask physicist Max Tegmark, who has written extensively about the implications of the multiverse, how he felt, he told me that he'd become skeptical about infinity. Me too, *mon frère*, me too!

35

Are We the Sim Sons (& Daughters)?

So, is there any way around the madhouse of infinity that lets us revive Jesus? Yes, there is. Perhaps, argues philosopher Nick Bostrom, what we think of as reality, with its immutable laws of nature, is actually a computer simulation of reality. Bostrom has an elaborate argument for why he thinks it is probable that we living in a simulation, but we need not go there. What's worth noting is that in a simulation what seem like unbreakable laws are really just programming choices that can be altered or bracketed by the programmer. Thus, all things really are possible with "God"—the programmer.

The physicist Frank Tipler makes a similar argument with an explicitly Christian twist. In his 1994 book *The Physics of Immortality*, Tipler lays out a tortuously complex argument whose essence is this: God lies in the future, arising from the expansion and deepening of intelligence in the universe, until, just in time to save us all from destruction, this techno-God simulates the entire history of the universe, with the resurrection of the dead thrown in for added thrills. It's an argument that is at once breathtaking and guffaw-inducing. (Full disclosure: about the same time, in complete ignorance of Tipler's work, I developed a secular argument that bears some resemblance to this. In 1997, the *New York Times* published a brief letter about it; a later, more fully developed version appears in my 2012 book *Free God Now!*)

Whether you take Bostrom's version or Tipler's, no rules that we can infer bind the programmer's choices. The creation stories offered in Genesis could be true down to the last detail. Every miracle may be true. As for the future, everything could change tomorrow, or the program could simply halt before tomorrow dawns. Prediction becomes a chump's game.

What's more, if we're in a simulation, we can't infer anything about the "real" universe from the appearance of ours (I owe this insight to Tegmark). So, whether God exists as an external programmer running the simulation that is our world (Bostrom), or emerges at the end of a long chain of evolution as the creator of the simulation that is our world (Tipler), we remain cloaked in ignorance about the ultimate reality.

You might think I've got that wrong in Tipler's case. After all, he argues that we're living or reliving the authentic history of the Universe. But there's a flaw in that apologetic ointment. Tipler's claim that the future godlike intelligence *will* simulate the past has an obvious corollary: it *may* also simulate alternative histories. Who's to say whether we exist in the real or alternative simulation?

In the end no simulation argument can satisfy our most basic epistemic need in relation to the Bible: to make sense of it. On a close reading, the epic story told by the Bible, culminating in the sacrifice of Jesus on the cross, is morally unintelligible. In the final chapter, the Book of Revelation, it becomes truly fantastical. Here, in just a short passage, is a four-cornered flat Earth, overarched by a canopy of tiny stars:

> I looked, and there came a great earthquake; the sun became black as sackcloth, the full moon became like blood, and the stars of the sky fell to the earth as the fig tree drops its winter fruit when shaken by a gale. The sky vanished like a scroll rolling itself up, and every mountain and island was removed from its place. Then the kings of the earth and the magnates and the generals and the rich and the powerful, and everyone, slave and free, hid in the caves and among the rocks of the mountains, calling to the mountains and rocks, "Fall on us and hide us from the face of the one seated on the throne and from the wrath of the Lamb; for the great day of their wrath has come, and who is able to stand?" After this I saw four angels standing at the four corners of the earth, holding back the four winds of the earth so that no wind could blow on earth or sea or against any tree. (Revelation 6:12–7:1, NRSV)

The entire Book of Revelation is littered with bizarre imagery—from horses with snake tails to a seven-headed scarlet beast that carries the "Whore of Babylon," who we are told is "clothed in purple and scarlet, and adorned with gold and jewels and pearls, holding in her hand a golden cup full of abominations and the impurities of her fornication."

Yuuuck! But not to worry, gentle readers, it all ends happily. Satan is captured, gets a thousand-year sentence, escapes, and immediately returns to his wicked ways. Another world war erupts, but this time Satan and all his followers draw a life sentence—well, an eternal sentence—in the lake of fire. God throws the old Earth onto the rubbish heap, creates a new world, complete with a new Jerusalem, moves in, and everything is clean, neat, and happy forevermore.

Welcome to Pleasantville.

36

 An Evo-Psych Solution

There's a good argument to be made that the Book of Revelation should be read as poetry, and many have made it. I would only add that if so, it is surely of the drug-induced hallucinatory genre that Samuel Taylor Coleridge, an opiate-swilling poet of Britain's Romantic period, did so much to promote.

But a peyote-style ending is not the only obstacle to making moral sense of the Bible. Consider a snappy summary: God creates humans with free will but requires them to follow his laws. Almost immediately, they fail to do so. It's evident he's created moral defectives. Rather than repair them, God curses one and kicks the pair out of paradise. They and their progeny continue to disappoint. God then drowns all of humanity and the animal kingdom—except one family. But that clan proves as morally defective as all the others: Noah's no sooner on dry land than he gets abominably drunk and curses his youngest son for no good reason. God then inflicts numerous punishments on his Chosen People, and deals out carnage, suffering, and death on their enemies. Finally, despairing of humanity's reform, God impregnates a woman and has the resulting son preach, prophesy, and perform miracles for several years before arranging for him to be cruelly sacrificed by his own people as "redemption" for their sins. Hmm . . .

The takeaway message? At the end of each gospel, it is not to be good, kind, compassionate, or wise, but rather to believe (Mark, John), to obey (Matthew), or simply to be forgiven (Luke).

This makes no moral sense. Indeed, take one step back and it becomes glaringly obvious that if God created the heavens and the earth, he could have just skipped the second part and stocked heaven with morally perfect creatures. Take another step back and something even more troubling becomes obvious: if God were infinitely powerful, knowing, and good, then he could not have created anything at all. To do so would either diminish the perfection already in place or indicate that God had not been maximally perfect to begin with. Perfection is not a variable.

Of course, reality has no obligation to satisfy our desire for logical consistency or moral intelligibility. But remember, science is the pursuit of the best-evidenced and most reliable knowledge we can obtain. Are we there yet? I think not.

No evidence backs either the infinite multiverse or the simulation argument. As speculation they allow for the biblical narrative, but they don't help us make sense of it. We are compelled, therefore, to consider another natural explanation: confabulation. Let's start with the climax of the Bible—the resurrection of Jesus. Could it have been sincerely *invented* as a way of making sense of disorienting events?

If the followers of Jesus believed him to be the Messiah, they could not have been expecting him to die on the cross, a broken and humiliated man. The Jews had definite ideas about what the Messiah was meant to be. As Rabbi Louis Jacobs writes in the Jewish Virtual Library:

> In rabbinic thought, the Messiah is the king who will redeem and rule Israel at the climax of human history and the instrument by which the kingdom of God will be established. . . . The Messiah was expected to attain for Israel the idyllic blessings of the prophets; he was to defeat the enemies of Israel, restore the people to the Land, reconcile them with God, and introduce a period of spiritual and physical bliss.

Things didn't work out that way. The Romans swatted Jesus like one more irritating fly on the rump of the Empire. Faced with a catastrophic reversal, his followers could only conclude that Jesus was yet one more false Messiah . . . or they could reinterpret events to make his death seem like part of the plan.

In his book *The Evolution of God*, Robert Wright explains:

In the gospels Jesus doesn't say he will return. He does refer to the future coming of a "Son of Man"—a term already applied in the Hebrew Bible to a figure who will descend from the skies at the climax of history. Yet Jesus never explicitly equates himself with the Son of Man. And in some cases he seems to be referring to someone other than himself. . . .

Jesus, like any good Jewish apocalyptic preacher of the time, affirmed apocalyptic scenarios in the Hebrew Bible, notably the one about the coming of the Son of Man. Then, after he died, his followers, stunned by the Crucifixion and trying to make sense of it, speculated that Jesus' references to the Son of Man had been veiled references to himself.

The key distinction here, lost in subsequent formulations of Christianity, is that after his death those followers who refused to give up on Jesus concluded that he was *not* the Messiah—he was something *more*. After all, the Messiah, as Tracey Rich, creator of the Web site Judaism 101, observes, was supposed to be a mighty but mortal king. A king like David—indeed, descended from David:

He will be well-versed in Jewish law, and observant of its commandments (Isaiah 11:2–5). He will be a charismatic leader, inspiring others to follow his example. He will be a great military leader, who will win battles for Israel. He will be a great judge, who makes righteous decisions (Jeremiah 33:15). But above all, he will be a human being, not a god, demi-god or other supernatural being.

When Jesus was nailed to the cross, with the mocking inscription "King of the Jews" tacked above his head, the possibility that he was the Messiah as that role had been understood perished with him. Wright speculates that Mary Magdalene and Mary the mother of Jesus may have had such a realization when they visited the tomb:

If the idea that Jesus was the Son of Man arose only after the Crucifixion . . . then the scene would have made perfect sense: the two Marys were having a major epiphany. Yes, they had heard that he might be the Messiah, the man who would save Israel, but they never dreamed that he could be something even greater: the Son of Man—the figure Jesus himself had so often glorified. (p. 309)

One could take this at theological face value. But our writ is to seek natural explanations, and nothing could be more natural than what Wright is suggesting: people, faced with a loss of heartfelt beliefs, often engage collectively in elaborate mythmaking so as to stave off a humiliating disillusionment. The stakes need not be all that high. Many Elvis Presley fans convinced themselves that "the King" was in hiding, or that he had been kidnapped, rather than face the unpalatable fact that he died on the toilet at age forty-two.

Religious figures have often been the target of that kind of wishful imagination. One version of Shi'a Islam holds that the Twelfth Imam, who was last heard from in 941 CE, is biding his time for a spectacular return *alongside* Jesus. Together, it is said, they will bring peace and justice to the world. Some Rastafarians believe that their hero, Emperor Haile Selassie of Ethiopia, who died in 1974, remains alive in seclusion.

I'm not equating Elvis with Jesus. One was a hip-shaking musician; the other arguably the most influential man in history. Rather, I'm pointing out that people demonstrably have a tendency to create narratives that sustain cherished beliefs.

These need not be conscious lies. We are not cameras. Perception and memory intermingle and weave stories out of selected data. Psychologists have identified numerous ways that we can be fooled: miscues, attentional overstimulation, misdirection, and suggestion, among others. And that's just for one person! When groups get to work on harmonizing their perceptions, mass delusion can arise.

At a young age I experienced this. Shortly after I started school, my class went outside for recess on a nice December day. Something, most likely a high-flying jet, went streaking overhead. One of my classmates yelled out, "It's Santa," and sure enough, I saw eight reindeer pulling Santa and his sleigh across the skies. So did all the other kids—unless some of them were lying. But even if they were, it made no difference. The group consensus was firm: we had seen Santa. I can close my eyes today and spot that sleigh.

Could the Resurrection be like that? Let's see. The stone that sealed the tomb was put in place by Joseph of Arimathea, so someone else could have rolled it away again. Angels, as ancient Israelites saw them, looked like other men. They had no wings, no harp, no golden halo.

Of course, the story of Pilate putting guards in front of the tomb because *everyone* knew Jesus had prophesied his resurrection in three days makes no sense. If that were true, his followers would have been camped out there by the hundreds or thousands waiting to see his triumphal reemergence. Instead, just a few women—the lowliest of people in the Bible—show up, and even those closest to Jesus are taken by surprise when they find the tomb empty. In short, no one was expecting the resurrection, and the story of the guards can only be a later invention in response to skeptical listeners. This means it is indeed possible that someone, or something, removed his body. Without resorting to the supernatural, many possibilities suggest themselves. The most palatable may be that one of his followers moved the corpse to another tomb out of fear the enemies of Jesus would desecrate it.

What of the sightings afterward? Clearly, in the natural world we know, Jesus could not have been literally dead one week and walk around showing off his wounds the next. But it's not hard to imagine that, with the tomb empty, rumors of sightings began. Again, think of the modern case of Elvis. No one has suggested that the popular singer was in any way a deity, and yet nearly forty years after his death, stories of Elvis sightings, some with video, continue to surface. (Google it for yourself.)

There's more. In an age when science and the scientific way of thinking are widely available, nearly a million Americans are passionately persuaded that they have been abducted by aliens and then returned to Earth. It's improbable enough that intelligent beings would cross the vast reaches of interstellar space to visit us, but any such technologically advanced beings would surely know what the polling company Gallup knows: that you can get a clear picture of a population by doing a random sample of a couple of thousand. A million? What a waste of energy!

What's more, if aliens were so intent on secrecy and so heedless of consent, they'd have no reason to expend more energy returning their victims alive. Yet, according to a 2012 survey by *National Geographic*, three-quarters of Americans believe that aliens have visited Earth. Now, that's credulity.

Considering the powerful incentives early Christians had to make themselves continue to believe and to persuade others, could they have jointly confabulated the story we know as the Resurrection? Of course they could. It's only natural.

If we view the Bible as a stream of ancient narratives created by people trying to make sense of the world, to inspire courage and unity in one another to face their many foes, and to constrain and channel the behavior of fellow tribalists for the benefit of the group, then it begins to make sense to us. The miracles, the commandments, and the narrative reframing of various catastrophes all point toward a common end: to promote the success of a group of people in ferocious competition for survival against hostile rivals and nature itself. Religion emerges as the binding force to overcome the selfish instincts of each individual and allow the group to triumph against the odds.

That's one scientific view of the function of religion in human societies. Darwin mused on it; evolutionary biologist and anthropologist David Sloan Wilson, in his book *Darwin's Cathedral*, makes the case for it. In a *Human Nature* review, philosopher Fritz Allhoff neatly summarizes the argument.

> Wilson argues that group selection has played an important role in evolutionary history and that, once we recognize this, much of the tension between religion and biology disappears. . . . [The] general idea is that selection can differentially favor groups whose members tend to possess certain traits, such as altruism. While altruism, by definition, diminishes the fitness of the possessor, a group full of altruists could be at a competitive advantage against a group of non-altruists. . . .
>
> If we introduce group selection, can we make headway on the tension between biology and religion? Wilson thinks that the answer is a resounding yes. Conceptually, the idea is that religion, which may impede certain individual reproductive interests, could nevertheless subserve the interests of groups. In other words, Wilson thinks that the tension between biology and religion will go away if natural selection is understood as a multi-level process such that group selection plays an important explanatory role. Much of *Darwin's Cathedral* . . . consists in Wilson's evidence for religion's promotion of group welfare. To this end, he considers many specific religions and, lest he be accused of

bias, also analyzes twenty-five religions chosen at random from the sixteen-volume Encyclopedia of World Religions. . . . How successful is Wilson's project? Clearly, he offers a lot of data in support of his view.

I've read the book, and I admit to an informed bias: I find it highly persuasive. The primary counterargument is that selfish individuals inside the group would overwhelm the altruistic population. But this fails to account for altruistic individuals banding together in mutual defense—as humans plainly do. In modern societies, we have justice systems and prisons for those who rob or rape. In ancient societies, there were social mechanisms—shaming, punishment, exile, religious laws, and threats of supernatural wrath—for misbehavior.

Even now, we don't always succeed at suppressing bad actors. Bullies and sociopaths are a recurring threat. The twentieth century taught us what tyrants armed with technology could do. Famine, genocide, and world wars followed.

Can science plausibly recast the Bible, with all its miracles and cruelties, its misogyny and moral crudity, as an instrument for cultivating altruism? Can any religion do that? Looking at ISIS, at gay-hating, misogynistic conservative Christians in America, or at brutal rightwing Buddhists in Myanmar, you might have trouble accepting this idea. All religions lend themselves to use as tools of oppression, aggression, and hate. Sadly, a call to love thy neighbor or turn the other cheek is no remedy.

Faith can't help being steeped in blood. For much of human existence, any society that wanted to survive had to inspire its young men to prepare for the ultimate sacrifice—to lay down their lives in a merciless war against other tribes. That's why the Bible includes passages like Psalm 137:7–9, which calls for a war of revenge against Babylon for its sacking of Jerusalem:

> Happy shall they be who pay you back what you have done to us! Happy shall they be who take your little ones and dash them against the rock!

That kind of Old Time Religion is still with us. The god of wrath, who inspired baby-killing verses like those quoted above, works his mischief in the minds of terrorists of all faiths. Yes, Islamic terrorists have a numerical lead and a gift for stealing headlines, but there are Christian terrorists, Jewish terrorists, Hindu terrorists, and Buddhist terrorists. Though they are out of fashion just now, there have been plenty of secular terrorists as well. The god of wrath can suit up in an abstract ideal (e.g., Marxism) just as well as in a religious one (Islamism, Reconstructionism, etc.).

Yet, for all the harms that religion can inflict in contemporary society, we must not lose sight of the beneficial function that it once had, and in some forms continues to have. In the absence of modern institutions, supernatural commandments were often the best we had—which is, admittedly, not much. When "Christendom" was at its height of power, murder rates in medieval Europe were more than ten times what they are now. Forensic science, coupled with modern law enforcement, is vastly more effective at curtailing bad behavior than religion ever was.

But to get to science, humankind needed social arrangements that fostered enough cooperation, wealth, literacy, specialization, and knowledge sharing to spark a revolution. Imagine how hard that was. For millions of years, our human ancestors lived in small, competing bands of hunter-gatherers. Anthropologist Robin Dunbar estimates the median size of our "tribes" was 150 people. Most of them would have been our close relatives. All others, be they Hittites, Perizzites, Girgashites, or Jebusites, were our enemies. It's likely that each band had its own dialect, making cross-tribal communication hard, and trust practically impossible. Skeletal evidence suggests that back in the day, murder lurked behind every bush. Rates of violent death hovered around 400 per 100,000—about a hundred times greater than the homicide rate in the United States today. Stranger danger had a level-up meaning.

Then, with the rise of agriculture, an evolutionary blink of an eye ago, mass societies began to pop up. But evolution has primed us to act in the interests of ourselves and our kinfolk, rather than to serve a vast tribe or state. The purpose of any individual's life, from an evolutionary point of view, is simple: to get as many copies of his or her genes as possible into the next generation. That imperative does not equip us to live in peace with our neighbors or to be kind to strangers.

Yet, humans are uniquely equipped to transcend our evolutionary imperatives and find better ways to live together and greater meaning in our lives. We do this largely through language. We tell stories that create fictive kinship bonds far greater than bounds of consanguinity. And those stories have largely been religious. Consider the words of Jesus: "You have heard that it was said, 'You shall love your neighbor and hate your enemy.' But I say to you, Love your enemies. . . . For if you love those who love you, what reward do you have? Do not even the tax collectors do the same?" (Matthew 5:44–46, NRSV) And here's a similar message in the Quran: "Remember the favor of Allah upon you, when you were enemies and he brought your hearts together and you became brothers by his favor." (Surah Ali Imran 3:103) "The believing men and believing women are allies of one another." (Surah at-Tawbah 9:71)

Of course, the Bible, Quran, and other scriptures also contain horrifying commandments (Pity the poor Perizzites!), and of course religion has been used as a tool of oppression, aggression, and enforced ignorance. Nevertheless, its prosocial functions in cultural evolution seem plain: religion initially cultivated altruism within tribal alliances, such as the ones that constituted ancient Israel, and then hit on the good trick of expanding the circle of kinship to all who adopted the faith.

Today, in advanced democracies, your chances of being murdered are vanishingly small—unless you are a minority living in an impoverished neighborhood. Even then, they are comparatively low. Worldwide, the rate of intentional homicide has dropped to just a tic over 6 per 100,000 people. How did we get from there to here?

Darwin recognized altruism as a hard problem in evolution. In the *Descent of Man*, he writes:

> It is extremely doubtful whether the offspring of the more sympathetic and benevolent parents, or of those which were the most faithful to their comrades, would be reared in greater number than the children of selfish and treacherous parents of the same tribe. He who was ready to sacrifice his life, as many a savage has been, rather than betray his comrades, would often leave no offspring to inherit his noble nature. The bravest men, who were always willing to come to the front in war,

and who freely risked their lives for others, would on an average perish in larger number than other men.

With David Sloan Wilson's idea in mind, let's look for an answer in the Bible. Here's an oft-quoted passage from John 15:13: "Greater love hath no man than this, that a man lay down his life for his friends." That's Jesus, lecturing his followers to be anti-evolutionary. Moments after inspiring them to altruistic self-sacrifice—the hard problem—he offers incentive: "whatsoever ye shall ask of the Father in my name, he may give it you." (John 15:16, KJV) That's some incentive, chum! Elsewhere in gospels, Jesus makes it clear that the package includes eternal life and unlimited resources.

Could the evolutionary implications be any clearer? At least in the instance of Christianity, religion functions to constrain selfish instincts for the benefit not just of kin but also of a faith-based fictive clan of unlimited size. Of course, it does not function optimally. From the Levites to the Medici popes, parasites have nearly always captured outsize benefits and manipulated religion to exploit or indeed to enslave others. Yet, a rational analysis has to conclude that for all its imperfections and contradictions, religion played a key role in setting the stage for science.

And, if religion has played a prosocial role in human affairs, then miracles have often served as stagehands. Christianity is the prime example. It is not only the largest religion, but miracles are key to its central story: the birth, death, and resurrection of Jesus Christ.

These conclusions may infuriate those who subscribe to the late Christopher Hitchens' view that religion poisons everything. Galileo, of course, was persecuted by the Catholic Church for mocking its dogmatic refusal to accept that the Earth goes around the Sun. Yet, from time immemorial, religion played a central role in governance, and if it had not been able to coerce, cajole, and inspire a measure of cooperation, the Scientific Revolution might not have taken place. Of course, that was never the goal of religion. The Church fathers had cause for concern that Galileo's telescopic findings would erode their credibility. His discovery of sunspots alone might have brought their Aristotelian Great Chain of Being crashing down. And if that weren't enough, Jupiter had its own surprises.

Huh. Whaddya know? There is no perfect realm beyond the orbit of the Moon. Instead, there are moons beyond the Moon! If Galileo was tough on the Vatican, think how devastating modern astronomy must be to the biblical description of the cosmos. Contrary to claims on the Internet, the Hubble Space Telescope has not spotted the Gates of Heaven. Instead, peering 13.4 billion light-years into space-time, it has documented a mind-boggling expanse clogged with at least 2 trillion galaxies. Remember, each galaxy contains billions of stars. If God created the universe with man in mind, he sure was wasteful.

But people don't easily change their minds. That's why miracles were needed in the first place. After Jesus was crucified, his remaining followers had an uphill battle to keep people interested in what otherwise must have looked like another failed entry in the Messiah sweepstakes. Bishop John Shelby Spong, in his book *A New Christianity for a New Age*, makes a persuasive case that miracles were piled onto the Jesus story long after his life. Spong notes that Paul, who wrote closest to the time of Jesus, has little to say of miracles other than the Resurrection. Spong also observes that many of the gospel miracles, including the Resurrection, appear to be recycled from earlier sources, both Jewish and pagan. In short, it looks like the longer Jesus failed to return, the more miracles were needed as part of a sales pitch to the skeptical.

Science has not been kind to miracles. Of course, people are free to go on believing that Noah built an ark and stuffed it with two of every kind—and many do. But, whereas that might have been easy to believe in the past, today it comes at a high price—and I'm not just talking about the Kentucky county that's facing bankruptcy after welcoming Ken Ham's Ark Encounter. To believe the tale of Noah's Ark, the faithful must willfully wall off both scientific evidence and their natural curiosity. They must not ask, for example, how a pair of lemurs got all the way from the Ark atop Ararat to the island of Madagascar, a journey of more than 4,000 miles (6,500 km), including more than 300 miles (500 km) of open ocean, leaving no offspring along the way. And if they do happen to ask, they must accept the flimsy rationalization that some apologist cooks up for them.

Science undermines both the credibility of miracles and their relevance to society. But this does not imply that religion, or secular analogues of the altruistic function it performs, will disappear. Quite the contrary.

The long arc of history clearly shows that societies in which people surrender some of their selfish interests for the commonwealth of cooperation flourish in comparison with those that embody the war of all against all or the bondage of tyranny. A few contemporary examples should suffice: compare the Czech Republic with Afghanistan, Costa Rica with Cuba, Denmark with Somalia, or, most strikingly, North Korea with South Korea.

After thousands of years as a nation, the Korean Peninsula was cleft in twain by the Cold War: the North went Stalinist under Kim Il Sung; the South became a military proxy state of the United States. However, in the South a gradual rise in the standard of living paved the way for a transition to democracy in 1987. Today, South Koreans enjoy one of the world's most advanced industrial economies and technological societies, while their kinfolk to the north live in isolated misery. Per capita income in the South is more than 16 times that in the North and life expectancy in the South is a decade longer than in the North. Most of all, however, people in the South have choices about their lives, while those in the North live to serve the whims of a dictator or face a hideous death.

In short, playing nice often pays. It opens the door to win-win transactions. As such, it is not pure altruism. And good thing, too. If we reconsider the story of Jesus on the cross, not even that emblem of sacrifice exhibits pure altruism. The human part of Jesus surely suffered. But the divine part of him always had to know that better times lay ahead. A return from the dead, a thousand-year rule, in heaven at the right hand of God—these are not inconsiderable rewards for sticking to the plan. Sacrifice for the common good can be a win-win game: that may be the contemporary gospel interpretation that serves us best. If miracles helped us get to a kinder, gentler, more scientifically informed world, perhaps we should pay our respects even as we inter them in the graveyard of dead ideas.

Yet, ironically, our scientific journey through the miracles of the Bible has beached us on a new River of Babylon. Within the scope of observable nature, we cannot validate biblical miracles as anything more than tricks, myths, or confabulations. But in the larger narrative of scientific speculation, we cannot dismiss them—at least not now.

As horrifying as it may seem, all logically possible stories may be true. Even the many contradictions of the Bible—from the different versions of the Creation story to the clashing accounts of the Gospels — may represent alternate versions existing in separate bubble universes, or different iterations of a simulation. But before anyone celebrates, consider that every possible version of the Bible's stories, including unwritten ones in which Satan triumphs, and Adam and Eve turn the Garden of Eden into a casino resort, would also have to be true. So would all the world's other religious narratives. Such speculations are completely useless when it comes to making sense of the Bible, or of life itself. Just as to form an image light needs dark, so too to create meaning truth needs falsity.

What can make sense of it all? Perhaps for an answer we need to look to life itself. If Wilson's right, evolution not only frames our beliefs and behavior, but also nudges us to create narratives that bend our selfish selves toward altruism. Can it be, then, that evolution, the scientific theory that best explains the web of life, also makes sense of the Bible's miracles? I believe it can.

Better still, informed by science of our evolutionary legacy as a hypersocial, often aggressive species with a growing capacity for self-destruction, such a view calls on us to reject or at least reinterpret tribalistic, supernatural narratives. An enlightened understanding of miracles instead encourages us to play nice with the rest of humanity. Love your neighbor as yourself—and recognize that we're all neighbors. Whether you choose to call that serving God or just plain humanism doesn't much matter in the end. Either way, it's a bit of a miracle, really.

ABOUT THE AUTHOR

Clay Farris Naff is a freelance science journalist and the *Humanist* magazine's science and religion correspondent. He is the author or editor of numerous books, including *Free God Now!*, and a variety of science readers. He lives in Lincoln, Nebraska.